AF360806

Symmetry in Renewable Energy and Power Systems

Symmetry in Renewable Energy and Power Systems

Editors

Raúl Baños Navarro
Alfredo Alcayde

MDPI • Basel • Beijing • Wuhan • Barcelona • Belgrade • Manchester • Tokyo • Cluj • Tianjin

Editors
Raúl Baños Navarro
Department of Engineering,
Electrical Engineering section,
University of Almería
Spain

Alfredo Alcayde
Department of Engineering,
Universidad de Almería
Spain

Editorial Office
MDPI
St. Alban-Anlage 66
4052 Basel, Switzerland

This is a reprint of articles from the Special Issue published online in the open access journal *Symmetry* (ISSN 2073-8994) (available at: https://www.mdpi.com/journal/symmetry/special_issues/Symmetry_Renewable_Energy_Power_Systems).

For citation purposes, cite each article independently as indicated on the article page online and as indicated below:

LastName, A.A.; LastName, B.B.; LastName, C.C. Article Title. *Journal Name* **Year**, *Volume Number*, Page Range.

ISBN 978-3-03943-901-0 (Hbk)
ISBN 978-3-03943-902-7 (PDF)

Cover image courtesy of Raúl Baños Navarro.

Contents

About the Editors

Raúl Baños Navarro was awarded his degree in Engineering (2001) and Ph.D. (2006) from the University of Almeria (Spain), and bachelor's degree in Economics from the National University of Distance Education (UNED). He is currently Lecturer at the Department of Engineering, University of Almeria. His research activities cover computational optimization, power systems, renewable energy, network analysis, and energy economics. As a result, he has published more than 150 papers in peer-reviewed journals, books, conferences, and workshops.

Alfredo Alcayde was awarded his bachelor's degree in Computer Science (2005) and Ph.D. (2011) from the University of Almeria (Spain) with the thesis "Optimisation in Electrical Engineering and Renewable Energies Using Evolutionary Multi-Objective Algorithms". He has published around 75 papers in journals, books, and conference proceedings.

Preface to "Symmetry in Renewable Energy and Power Systems"

The study of power systems is closely related with symmetry. For example, multiphase power systems are inherently symmetric. The study of symmetrical and asymmetrical faults in power systems is a critical issue. The phase sequence arrangements of multicircuit overhead lines on the same tower directly affects the symmetry of power transmission systems, which influences the operation of power grid and relay protection. Moreover, symmetry is a topic of intensive investigation in the analysis of grid interconnection, including symmetrical and asymmetrical network parameters in smart grid infrastructures. In renewable energy, the symmetry is present in the layout of wind power plants or photovoltaic plants, among others, while the performance of solar systems can differ if they are used with symmetric and asymmetric concentrating CPC collectors. Therefore, investigations related to symmetry in power systems and renewable energy will favor the advance in these disciplines.

Raúl Baños Navarro, Alfredo Alcayde
Editors

Article

Second-Order Cone Approximation for Voltage Stability Analysis in Direct-Current Networks

Oscar Danilo Montoya [1,2] **and Walter Gil-González** [3] **and Alexander Molina-Cabrera** [4,*]

[1] Facultad de Ingeniería, Universidad Distrital Francisco José de Caldas, Bogotá 11021, Colombia; odmontoyag@udistrital.edu.co or omontoya@utb.edu.co

[2] Laboratorio Inteligente de Energía, Universidad Tecnológica de Bolívar, Cartagena 131001, Colombia

[3] Grupo GIIEN, Facultad de Ingeniería, Institución Universitaria Pascual Bravo, Campus Robledo, Medellín 050036, Colombia; walter.gil@pascualbravo.edu.co

[4] Facultad de Ingeniería, Universidad Tecnológica de Pereira, Pereira 660003, Colombia

* Correspondence: almo@utp.edu.co

Received: 16 August 2020; Accepted: 22 September 2020; Published: 24 September 2020

Abstract: In this study, the voltage stability margin for direct current (DC) networks in the presence of constant power loads is analyzed using a proposed convex mathematical reformulation. This convex model is developed by employing a second-order cone programming (SOCP) optimization that transforms the non-linear non-convex original formulation by reformulating the power balance constraint. The main advantage of the SOCP model is that the optimal global solution of a problem can be obtained by transforming hyperbolic constraints into norm constraints. Two test systems are considered to validate the proposed SOCP model. Both systems have been reported in specialized literature with 6 and 69 nodes. Three comparative methods are considered: (a) the Newton-Raphson approximation based on the determinants of the Jacobian matrices, (b) semidefinite programming models, and (c) the exact non-linear formulation. All the numerical simulations are conducted using the MATLAB and GAMS software. The effectiveness of the proposed SOCP model in addressing the voltage stability problem in DC grids is verified by comparing the objective function values and processing time.

Keywords: convex reformulation; direct current networks; non-linear optimization; numerical example; second-order cone programming; voltage stability margin

---

1. Introduction

Direct current (DC) electrical networks are promising grids capable of supplying multiple users at different voltage levels from high-voltage DC (HVDC) to low-voltage DC (LVDC) in monopole or bipole configurations [1,2]. The implementation of DC technologies avoids the need for managing reactive power or frequency, in contrast to their alternating current (AC) counterparts. This is an important advantage that makes DC grids easily controllable and operable. Additionally, power losses are lower, and voltage profiles are better in DC grids than in AC systems. Hence, DC networks are more efficient than AC networks [3,4].

Two types of strategies are used to analyze DC electrical networks: dynamical and static approaches. The first strategy is executed in the time domain and is used for developing primary and secondary controllers in power electronic DC-DC converters [5,6]. The second type of analysis, i.e., static studies, is used to determine all the state variables under stationary conditions. The most typical types of analysis are power flow analysis [3], optimal power flow studies [2], economic dispatch approaches [7], and voltage stability analysis [8,9]. In addition, these approaches are combined with the optimal sizing and location of distributed generators for DC grids [10].

In this study, we focus on the voltage stability calculation for DC grids. This is a non-linear non-convex optimization problem recently analyzed in specialized literature, and few approaches for this task have been reported. In [9], a semidefinite programming (SDP) model was proposed by guaranteeing a global and unique solution. Nevertheless, the complexity of this model is mainly due to the quadratic increase in the number of variables and the semidefinite requirements of the matrix that contains all the voltage variables; this causes longer computational times when the number of nodes in the DC system increases. The authors of [8] presented a linear matrix inequalities formulation to determine the maximum load increments in a small DC grid composed of two constant-power loads. However, this approach cannot be extended for multiple loads, because the resulting equations are unsolvable using analytical methods. The authors of [11] employed the classical Newton–Raphson method in conjunction with a linear search to determine the maximum load increment. This is performed by observing the sign variation in the Jacobian matrix in the power flow equations. This method is easily implementable; however, the selection of the step size has an undesirable influence on the final solution. In [12], the voltage stability margin problem in DC grids was solved by incorporating the non-linear formulation into an optimization package known as the general algebraic modeling system (GAMS). Even though this software is efficient for solving non-linear problems, it is not possible to guarantee a global solution to the problem, because the calculations are usually stuck in local solutions.

Unlike in previous works, in this study, a second-order cone programming (SOCP) model is proposed to address the voltage stability margin calculation in DC grids. The main advantage of this approach is that it guarantees a global optimum and unique solution by transforming the exact non-linear non-convex optimization problem into a convex problem [13,14]. In addition, this approach has not been previously proposed for the analysis of voltage stability in DC networks. Therefore, there is a gap in the literature that this study tries to fulfill. The convex approach has lower computational requirements than SDP approaches because it avoids semidefinite matrices in its formulation.

Even if recently reported approaches typically use the exact non-linear formulation (GAMS solvers) and heuristic searches (Newton-Raphson) [11,12] because of the non-convexities introduced by the power balance constraints, it is not possible to ensure a global optimum, even if for both test feeders these solutions coincide with convex approaches, i.e., the semidefinite programming model [9] and the newly proposed SOCP model.

The remainder of this paper is organized as follows: Section 2 presents the classical non-linear non-convex formulation of the voltage stability calculation in DC grids. Section 3 shows the proposed second-order cone programming reformulation and its main assumptions for developing a convex mathematical model. Section 4 presents a small numerical example with three nodes to demonstrate the effect of the load increment and the voltage collapse problem. Section 5 shows the numerical implementation of the proposed SOCP model in two test systems, namely an HVDC network and a medium-voltage DC (MVDC) grid. Section 6 presents the main conclusions drawn from this research.

2. Non-Linear Programming Formulation

The determination of the point of the voltage collapse in electrical DC networks with constant-power loads is formulated as a non-linear non-convex optimization problem [9]. The non-convexities of this problem are related to the power balance equations in the presence of constant-power loads, as these expressions emerge as a hyperbolic relation between voltage and power that generates non-affine equality constraints [8]. The complete optimization model for analyzing the point of voltage collapse in DC grids is formulated as follows:

Objective function:

$$\max z = \sum_{i=1}^{n} (1 + \lambda_i)\, p_i^d, \tag{1}$$

where z is the value of the objective function related to the maximum power consumption possible in the DC grid; λ_i represents the decision variable associated with the increment in the constant-power load at each node; and p_i^d is the constant power consumption connected to node i. Please note that n is the total number of nodes in the DC grid. In addition, the objective function (1) is linear, which makes it convex in the solution space.

Set of constraints:

$$p_i^g - (1 + \lambda_i)\, p_i^d = v_i \sum_{j=1}^n g_{ij} v_j, \; i \in \mathcal{N}, \tag{2}$$

$$v_i^{\min} \leq v_i \leq v_i^{\max}, \; i \in \mathcal{N}, \tag{3}$$

$$p_i^{g,\min} \leq p_i^g \leq p_i^{g,\max}, \; i \in \mathcal{N}, \tag{4}$$

$$\lambda_i \geq 0, \; i \in \mathcal{N} \tag{5}$$

where p_i^g represents the power generation by the voltage-controlled nodes (i.e., slack nodes) connected to node i; g_{ij} corresponds to the conductance effect that relates nodes i and j and is considered to be a constant parameter that depends on the node interconnections. v_i and v_j are the voltage variables associated with nodes i and j, respectively, and are lower- and upper-bounded by $v_i^{\min}$ and $v_i^{\max}$, respectively. Finally, $p_i^{g,\min}$ and $p_i^{g,\max}$ correspond to the minimum and maximum power generation bounds in the slack nodes, respectively.

Remark 1. *Please note that the decision variables in the problem of determining the stability margin in DC networks shown in (1)–(5) correspond to the loadability parameter λ_i as well as the voltage profiles in all the nodes, i.e., v_i and the power generation in the constant-voltage sources. This suggests that the solution of this problem involves the simultaneous determination of all these variables, which maximizes the chargeability of the grid [12].*

The optimization model defined in (1)–(5) receives the following interpretation: Equation (1) is the objective function that corresponds to the maximization of the total power consumption admissible in all the nodes immediately before reaching the point of the voltage collapse. Equation (2) is the power balance constraint. Equation (3) defines the power capabilities in all generation nodes. Equation (4) determines the voltage bounds admissible for secure operation of the DC grid under normal operative scenarios, and Equation (5) determines the positiveness nature of the loadability variable.

Remark 2. *If the variable λ_i is zero for all the nodes, then the mathematical model (1)–(5) corresponds to the classical power flow problem for DC grids, which can be solved using classical methods, such as the Gauss–Seidel [2], Newton–Raphson [3] or successive approximation methods [15], among others. All these approaches can guarantee the existence and uniqueness of the solution under well-defined voltage conditions through fixed-point theorems.*

Please note that the objective of voltage stability analysis is to determine the maximum power increments in all the constant power loads that carry the DC system to the voltage collapse. Hence, the constraints related to generation capabilities in slack nodes, and voltage bounds in all the nodes are relaxed [12]. Therefore, these constraints are neglected when the objective of the problem is to determine the point at which all the nodes have a voltage collapse [11].

Remark 3. *Even though non-linear optimization methods such as the interior-point or gradient-descent methods can solve model (1)–(5), there is no guarantee of reaching the global optimum because of the non-convexity of the power balance constraints.*

The mathematical problem (1)–(5) is transformed into a convex one using semidefinite programming or SOCP to guarantee the uniqueness of the mathematical solution of the voltage stability margin determination in DC grids [9,16]. We used the SOCP to solve the voltage stability determination problem, which represents the main contribution of this study. The SOCP model is described in the next section.

Convexity Test

To demonstrate that the power balance equations in DC networks represent a set of non-linear non-convex constraints, we present a small numerical example as follows: consider a DC power system composed of 3 nodes (see the test feeder presented in Section 4), i.e., one voltage-controlled node and two constant-power loads. For this example, let us rewrite the power balance equation at node 2 using the per-unit (p.u.) representation.

$$-p_{d_2} = G_{20}v_2v_0 + G_{21}v_1v_2 + G_{22}v_2^2, \tag{6}$$

for simplicity, let us consider that $p_{d_2} = 1/4$ p.u, $v_0 = 1$ p.u, $G_{20} = -1/2$ p.u, $G_{21} = -1/2$ p.u, $G_{22} = 1$ p.u, $v_1 = x$, and $v_2 = y$, which produces:

$$2y^2 - xy - y + \frac{1}{4} = 0. \tag{7}$$

If we plot the non-linear function (7), the curve illustrated in Figure 1 is obtained. Please note that the solution space is given by the red curve, which implies that it is convex only if a linear combination $tx_1 + (1-t)x_2 = x$ is contained in the curve, where t is a real number between 0 and 1. In Figure 1, it can be observed that the line points generated (blue line) by the linear combination are outside the red curve (except the extreme points). This means that the constraint (7) is non-convex. Hence, the power balance constraints in power flow analysis generate a non-convex solution space. This implies that it is impossible to ensure a global optimum in power flow analysis.

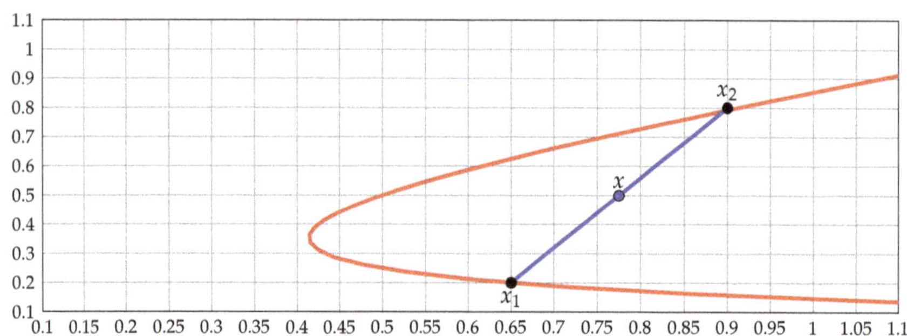

Figure 1. Numerical test to show the non-convexity of the power balance equations.

3. Second-Order Cone Programming Formulation

The SOCP formulation is a component of optimization convex models. This is an approach that has gained considerable importance in engineering because it can solve a family of convex problems reliably and efficiently by guaranteeing a unique solution (global optimum) [17]. The SOCP formulation minimizes a linear function over a convex region, which consists of the intersection of second-order cones with an affine linear space [18].

To transform the problem of the voltage stability margin (formulated from (1)–(5)) into an SOCP model, it is important to mention that the only one non-convex constraint represents the power balance

Equation (2). To perform this transformation, we focus on the product between voltage variables, i.e., $v_i v_j$, by redefining a new variable y as follows:

$$y_{ij} = v_i v_j,,$$ (8)

Now, if we multiply (8) by y_{ij}, then, the following result yields

$$y_{ij}^2 = v_i^2 v_j^2 \leftrightarrow ||y_{ij}||^2 = ||v_i||^2 ||v_j||^2,$$ (9)

where the pre-multiplication by $v_i v_j$ is required to transform the hyperbolic relation between voltages into a conic constraint [19].

Please note that this relaxation is possible because all the voltage variables must be positive for satisfactory operation of DC grid, including in extreme cases, such as the voltage stability margin analyzed in this study.

In (9), it is possible to substitute expression (8), which yields the following result

$$||y_{ij}||^2 = y_{ii} y_{jj}.$$ (10)

Please note that (10) is still non-linear non-convex, which implies the need for a relaxation. The first step is to relax the equality constraint using an inequity as follows:

$$||y_{ij}||^2 \leq y_{ii} y_{jj}.$$ (11)

Remark 4. *Please note that the relaxation of the equality imposition by a lower-equality imposition is required at any conic approximation because equality implies that the solution is only in the contour of the cone. The lower-equal symbol implies that all the points inside the cone are possible solutions, including the contour of the cone, which implies that this relaxation passes the convexity test presented in Section 2 [20].*

Theorem 1. *Hyperbolic constraint (11) can be transformed into a conic constraint as follows:*

$$\left\| \begin{array}{c} 2y_{ij} \\ y_{ii} - y_{jj}, \end{array} \right\| \leq y_{ii} + y_{jj}.$$ (12)

Proof. Let us elevate to square both sides of the expression (12), which yields

$$\left\| \begin{array}{c} 2y_{ij} \\ y_{ii} - y_{jj}, \end{array} \right\|^2 \leq (y_{ii} + y_{jj})^2,$$ (13)

This expression can be rewritten as follows:

$$\left\| \begin{array}{c} 2y_{ij} \\ y_{ii} - y_{jj}, \end{array} \right\|^T \left\| \begin{array}{c} 2y_{ij} \\ y_{ii} - y_{jj}, \end{array} \right\| \leq (y_{ii} + y_{jj})^2.$$ (14)

Now, if we expand all the components in (14), then the following result is obtained:

$$4y_{ij}^2 + y_{ii}^2 - 2y_{ii} y_{jj} + y_{jj}^2 \leq y_{ii}^2 + 2y_{ii} y_{jj} + y_{jj}^2,$$
$$4y_{ij}^2 \leq 4y_{ii} y_{jj}$$
$$y_{ij}^2 \leq y_{ii} y_{jj},$$ (15)

Please note that (15) is the same as to (11), and the proof is complete. $\square$

Because the power flow constraint can be rewritten as a set of convex restrictions, the equivalent SOCP model representing the problem of maximum loadability in DC networks with constant-power loads can be rewritten as follows:

Objective function:

$$\max z = \sum_{i=1}^{n} (1 + \lambda_i)\, p_i^d,\tag{16}$$

Set of constraints:

$$p_i^g - (1 + \lambda_i)\, p_i^d = \sum_{j=1}^{n} g_{ij} y_{ij}, \ i \in \mathcal{N},\tag{17}$$

$$\left\|\begin{array}{c} 2y_{ij} \\ y_{ii} - y_{jj}, \end{array}\right\| \le y_{ii} + y_{jj}, \ i, j \in \mathcal{N}\tag{18}$$

$$v_i^{\min} v_j^{\min} \le y_{ij} \le v_i^{\max} v_j^{\max}, \ i, j \in \mathcal{N},\tag{19}$$

$$p_i^{g,\min} \le p_i^g \le p_i^{g,\max}, \ i \in \mathcal{N},\tag{20}$$

$$\lambda_i \ge 0, \ i \in \mathcal{N}\tag{21}$$

Remark 5. *Mathematical models (1)–(5) and (17)–(21) are equivalent in (18) if it is guaranteed that the quality characteristic will be maintained in (19).*

Remark 6. *To retrieve the original optimization variables in the SOCP model described in (17) to (21), the following expression can be used:*

$$v_i = \sqrt{y_{ii}}, \ i \in \mathcal{N},\tag{22}$$

4. Graphical Example

Here, we considered a small DC test feeder composed of three nodes (one of them is the slack node) and two constant-power loads connected to nodes 1 and 2, respectively. This system is used to show the voltage stability margin, i.e., the region of secure operation, graphically. The topology of this test feeder is presented in Figure 2. For this test system, 1000 W and 24 V are considered the power and voltage bases, respectively.

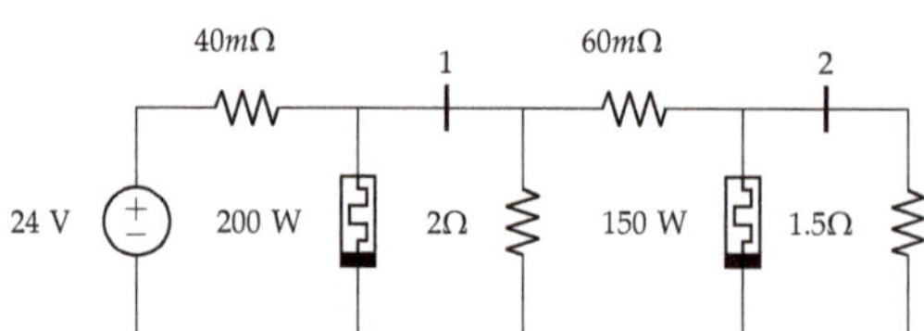

Figure 2. Small test feeder with two constant power loads.

Figure 3 shows the numerical behavior of the voltage collapse point when different increments have been used at the point of connection of the constant-power loads. Please note that point **O** is the solution of the classical power flow problem when all λ_i are fixed as zero; this point is $(v_1, v_2) = (0.9312, 0.8783)$. From this initial point, we evaluate the evolution of the voltage collapse in the DC grid when its constant-power loads increase. Trajectory **O–A** shows the evolution of the voltage at load nodes when it is increased only at the load connected at node 1, with the load at node 2 being fixed as 0.15 p.u. Please note that point **A** presents the maximum reduction in both load voltages at the same time, i.e., $(v_1, v_2) = (0.4802, 0.4265)$; both voltages are observed to be lower than 0.5000 p.u.

In addition, these points represent the maximum objective function possible in this numerical example, which is $z = 3.4304$ p.u, when $\lambda_1 = 15.4021$. Trajectory **O–B** shows the evolution of the voltage in the DC system when both loads are increased by the same magnitude, i.e., $\lambda_1 = \lambda_2 = 5.5319$. These increments in the loads produce a maximum objective function of 2.2862 p.u, where one voltage is higher than 0.6500 p.u (see node 2), and the other node is lower than 0.4500 p.u (see node 3). This behavior implies that node 3 conditioned the stability margin behavior of this test system since it is more sensitive to load changes than node 2 is. On the other hand, trajectory **O–C** presents the voltage evolution of the numerical example when the load connected to node 2 is increased, with the load at node 1 fixed as 0.20 p.u. This trajectory shows that the voltage at node 2 decreases until 0.4534, while the voltage at node 1 remains upper that 0.7600 p.u. This behavior confirms that node 2 has a lower possibility of increasing its load consumption since the voltage collapse point is reached when the total load of the DC system is 1.4611 p.u, which is the minimum objective functions across the three cases analyzed. Please note that in Table 1 the numerical behavior of the voltage stability problem in DC networks resumes when constant-power loads start to increase.

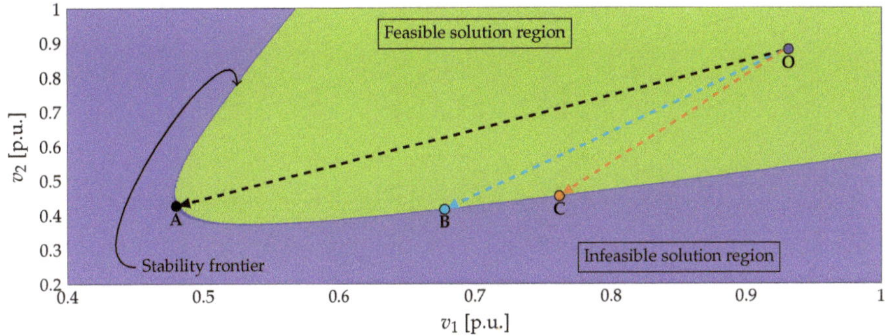

Figure 3. Voltage collapse trajectories followed by different increments in the constant power consumptions per node.

Table 1. CPL increments for the different simulation cases.

Trajectory	λ_1	λ_2	Collapse Point (v_1, v_2)	z [p.u.]
O–A	15.4021	0	A (0.4802,0.4265)	3.4304
O–B	5.5319	5.5319	B (0.6776,0.4151)	2.2862
O–C	0	7.4076	C (0.7613,0.4534)	1.4611

It is important to mention that the results presented in this numerical example have differences lower than 1 % when compared with that in the heuristic approach based on Newton–Raphson sensitivities [11].

Remark 7. *The point of voltage collapse in DC radial networks depends on the location of the load node. Hence, it is possible to conclude from the numerical example that loads connected to the final nodes have lower possibilities of incrementing their consumption when compared with loads near the slack source.*

5. Test Systems and Simulation Results

In this section, we present the test system configuration and the numerical results obtained by solving the stability margin calculation problem with different methodologies reported in the specialized literature.

5.1. Test System Configurations

To validate the proposed SOCP model for voltage stability calculations in DC networks, we consider the testing feeders reported in [11]. The first DC network corresponds to the high-voltage DC (HVDC) network, and the second network is an MVDC radial network. The topologies of the test feeders are illustrated in Figure 4. All the numerical information related to load consumption and branch parameters can be obtained from [11].

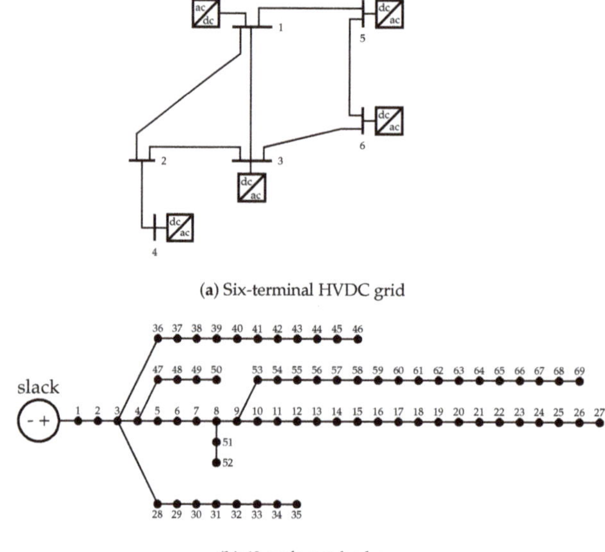

(**a**) Six-terminal HVDC grid

(**b**) 69-node test feeder

Figure 4. Electrical configuration of the test systems.

5.2. Numerical Validation

The proposed SOCP formulation was validated by being compared with approaches reported in the specialized literature. The interior point was used for solving the exact non-linear programming formulation (IP-NLP), the Newton–Raphson formulation based on determinants of Jacobian matrices (NR-DJM), and SDP formulations [9,11]. These methodologies were implemented in MATLAB and GAMS.

Table 2 presents the maximum loadability factor, i.e., λ, for the HVDC and MVDC systems. Please note that the proposed SOCP model allows for reaching the optimal global solution of this problem for both test systems since it is a convex transformation of the exact non-linear non-convex problem.

Table 2. Voltage stability index for the HVDC and MVDC test feeders.

Test System	NR-DJM	IP-NLP	SDP	SOCP
HVDC	5.6588	5.6588	5.6588	5.6588
MVDC	3.0200	3.0200	3.0067	3.0200

In the case of the HVDC system, the proposed solution technique, i.e., the SOCP model, as well as the comparative methods, reach the same loadability factor ($\lambda = 5.6588$) for all the nodes. In contrast, for the MVDC system, the SDP model presents an underestimation of the loadability factor when compared with the global optimum, i.e., 3.0200. This error is approximately 0.44% when the SDP is compared with NR-DJM, IP-NLP, and the proposed approach.

In terms of the computational performance with respect to the processing times required to solve the voltage stability margin problem, all the methodologies listed in Table 2 require between 0.25 s and 30 s in the case of SOCP. In the case of the proposed approach, for the HVDC system, the processing time is 0.28 s, while for the MVDC, the time is 4.611 s. These results confirm that the proposed SOCP model is more efficient when compared with the NR-DJM (4.19 s and 21.11 s), SDP (0.30 s and 12.44 s), and the IP-NLP (0.28 s and 3.57 s) models reported in [11].

In the case of the NR-DJM, it is difficult to select the heuristic parameter α reported in [11] to determine the convergence of the algorithm. In case of the 69-node test feeder, before the voltage collapse ($\lambda = 3.01$), the DJM is approximately 3.1856×10^{268}, and at the point of the voltage-collapse, it is approximately -7.9939×10^{265}. This implies that the tuning of this heuristic search requires multiple power flow evaluations. An additional complication of the NR-DJM approach is the selection of the step δ, because large values of this parameter make the algorithm faster but sacrifice precision, while small values increase the precision of the method. Large values also increase the computational time required for the solution of the problem. In other words, even if the NR-DJM method is intuitive and easy to implement, it requires adequate parametrization of the algorithm, which makes it highly dependent on the programmer. However, this is not the case with the proposed SOCP approach; this approach does not require any adjustment parameter.

Figure 5 depicts the voltage profile for the power flow problem considering all the chargeability factors as zero and the voltage collapse point when all the constant-power loads increase in the same magnitude.

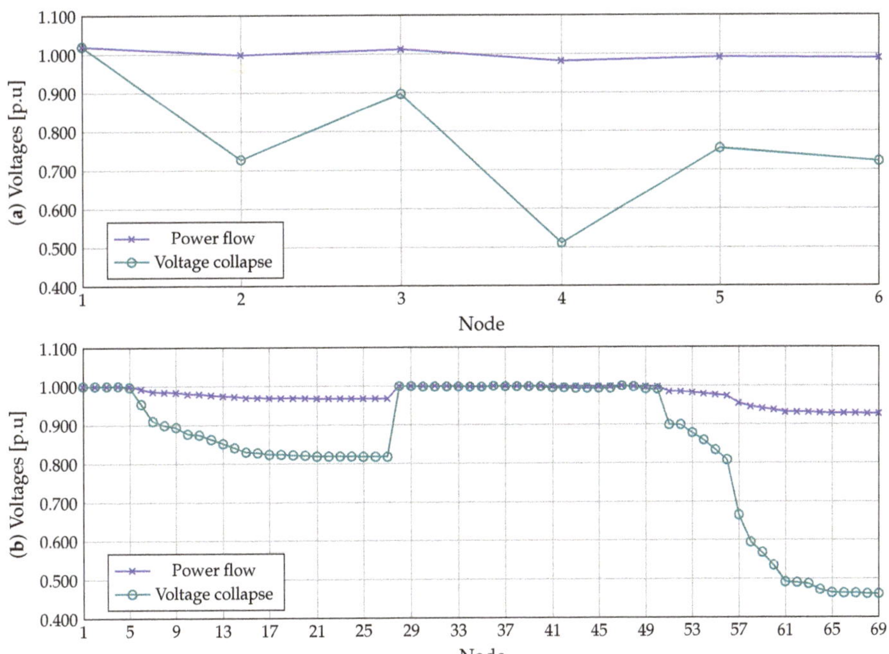

Figure 5. Voltage behavior in the test system for the initial state of load and voltage collapse: (**a**) HVDC test system, and (**b**) MVDC test feeder.

From Figure 5, it is possible to extract the following behaviors:

✓ The HVDC test feeder with its meshed structure maintains voltages higher than 0.50 p.u. until the voltage collapse scenario. Please note that node 4 presents the lower voltage profile with 0.5114 p.u., which is a radial extension of this HVDC system.

✓ The voltage collapse in the MVDC test feeder is evident long after node 57 and onward. This situation occurs in this part (node 57 and onward) of the test feeder since the total load is more significant than regarding routes. Voltage collapse occurs when the maximum voltage drop is 0.4610 p.u. at node 69.

✓ In both test systems, the voltage collapse occurs when voltages are lower than 0.55 p.u; while the total load consumptions increase at least three times. This behavior implies that the power system protection disconnects this system before the voltage collapse occurs because of the high currents flowing through the branches.

6. Conclusions and Future Work

A convex reformulation of the voltage stability margin determination in DC networks with constant-power loads was proposed in this paper based on a second-order cone formulation. This model guarantees a global optimum for the problem by approximating the hyperbolic constraints related to the power flow problem. Numerical results, in comparison with those of the Newton–Raphson, SDP, and interior-point approaches, show the efficiency of the proposed SOCP formulation in terms of the objective function calculation and processing times required.

An analysis of voltage stability margin in DC grids enabled the determination of the maximum range of load increments before the collapse of the network. This is important because utilities can use these results for planning and grid reposition procedures (changes in the size of conductors, substations, grid topology) to avoid blackouts when new users are interconnected.

In future work, it will be possible to extend the convex SOCP formulation presented in this paper for optimal power flow analysis, i.e., economic dispatch, in DC networks considering daily operation with a high penetration of renewable energy resources and batteries.

Author Contributions: Conceptualization, O.D.M., W.G.-G., and A.M.-C.; methodology, O.D.M., W.G.-G., and A.M.-C.; formal analysis, O.D.M., W.G.-G., and A.M.-C.; investigation, O.D.M., W.G.-G., and A.M.-C.; resources, O.D.M., W.G.-G., and A.M.-C.; writing—original draft preparation, O.D.M., W.G.-G., and A.M.-C. All authors have read and agreed to the published version of the manuscript. These authors contributed equally to this work.

Funding: This work was partially supported by the National Scholarship Program Doctorates of the Administrative Department of Science, Technology, and Innovation of Colombia (COLCIENCIAS), by calling contest 727-2015.

Acknowledgments: The authors want to thank Vicerrectoria de Investigación, Innovación y Extensión from Universidad Tecnológica de Pereira for the support provided in this investigation.

References

1. Dragičević, T.; Lu, X.; Vasquez, J.C.; Guerrero, J.M. DC microgrids–Part I: A review of control strategies and stabilization techniques. *IEEE Trans. Power Electron.* **2016**, *31*, 4876–4891.

2. Garces, A. Uniqueness of the power flow solutions in low voltage direct current grids. *Electr. Power Syst. Res.* **2017**, *151*, 149–153. [CrossRef]

3. Garcés, A. On the Convergence of Newton's Method in Power Flow Studies for DC Microgrids. *IEEE Trans. Power Syst.* **2018**, *33*, 5770–5777.

4. Gan, L.; Low, S.H. Optimal power flow in direct current networks. *IEEE Trans. Power Syst.* **2014**, *29*, 2892–2904. [CrossRef]

5. Rouzbehi, K.; Miranian, A.; Candela, J.I.; Luna, A.; Rodriguez, P. A Generalized Voltage Droop Strategy for Control of Multiterminal DC Grids. *IEEE Trans. Ind. Appl.* **2015**, *51*, 607–618. [CrossRef]

6. Rouzbehi, K.; Miranian, A.; Luna, A.; Rodriguez, P. DC Voltage Control and Power Sharing in Multiterminal DC Grids Based on Optimal DC Power Flow and Voltage-Droop Strategy. *IEEE J. Emerg. Sel. Top. Power Electron.* **2014**, *2*, 1171–1180, [CrossRef]

7. Hamad, A.A.; El-Saadany, E.F. Multi-agent supervisory control for optimal economic dispatch in DC microgrids. *Sustain. Cities Soc.* **2016**, *27*, 129–136. [CrossRef]

8. Barabanov, N.; Ortega, R.; Griñó, R.; Polyak, B. On Existence and Stability of Equilibria of Linear Time-Invariant Systems With Constant Power Loads. *IEEE Trans. Circuits Syst. I* **2016**, *63*, 114–121, [CrossRef]

9. Montoya, O.D. Numerical Approximation of the Maximum Power Consumption in DC-MGs With CPLs via an SDP Model. *IEEE Trans. Circuits Syst. II* **2019**, *66*, 642–646, [CrossRef]

10. Grisales-Noreña, L.F.; Gonzalez Montoya, D.; Ramos-Paja, C.A. Optimal sizing and location of distributed generators based on PBIL and PSO techniques. *Energies* **2018**, *11*, 1018. [CrossRef]

11. Montoya, O.D.; Gil-González, W.; Garrido, V.M. Voltage Stability Margin in DC Grids with CPLs: A Recursive Newton–Raphson Approximation. *IEEE Trans. Circuits Syst. II* **2019**, *67*, 300–304. [CrossRef]

12. Grisales-Noreña, L.F.; Garzon-Rivera, O.D.; Montoya, O.D.; Ramos-Paja, C.A. Hybrid Metaheuristic Optimization Methods for Optimal Location and Sizing DGs in DC Networks. In *Workshop on Engineering Applications*; Chapter Applied Computer Sciences in Engineering; Figueroa-García, J., Duarte-González, M., Jaramillo-Isaza, S., Orjuela-Cañon, A., Díaz-Gutierrez, Y., Eds.; Springer: Berlin, Germany, 2019; Volume 1052, pp. 552–564. [CrossRef]

13. Li, J.; Liu, F.; Wang, Z.; Low, S.H.; Mei, S. Optimal Power Flow in Stand-Alone DC Microgrids. *IEEE Trans. Power Syst.* **2018**, *33*, 5496–5506. [CrossRef]

14. Xie, Y.; Chen, X.; Wu, Q.; Zhou, Q. Second-order conic programming model for load restoration considering uncertainty of load increment based on information gap decision theory. *Int. J. Electr. Power Energy Syst.* **2019**, *105*, 151–158. [CrossRef]

15. Montoya, O.D.; Garrido, V.M.; Gil-González, W.; Grisales-Noreña, L.F. Power Flow Analysis in DC Grids: Two Alternative Numerical Methods. *IEEE Trans. Circuits Syst. II* **2019**, *66*, 1865–1869. [CrossRef]

16. Lavaei, J.; Low, S.H. Zero Duality Gap in Optimal Power Flow Problem. *IEEE Trans. Power Syst.* **2012**, *27*, 92–107. [CrossRef]

17. Hindi, H. A tutorial on convex optimization. In Proceedings of the 2004 American Control Conference, Boston, MA, USA, 30 June–2 July 2004; Volume 4, pp. 3252–3265.

18. Alizadeh, F.; Goldfarb, D. Second-order cone programming. *Math Program* **2003**, *95*, 3–51. [CrossRef]

19. Yuan, Z.; Hesamzadeh, M.R. Second-order cone AC optimal power flow: Convex relaxations and feasible solutions. *J. Mod. Power Syst. Clean Energy* **2018**, *7*, 268–280. [CrossRef]

20. Luo, Z.Q.; Yu, W. An introduction to convex optimization for communications and signal processing. *IEEE J. Sel. Areas Commun.* **2006**, *24*, 1426–1438. [CrossRef]

Article

Coordinated Control and Dynamic Optimal Dispatch of Islanded Microgrid System Based on GWO

Yuting Wang *, Chunhua Li and Kang Yang

Electronic and Information College, Jiangsu University of Science and Technology, Zhenjiang 212003, China; viven_lch@just.edu.cn (C.L.); 192030035@stu.just.edu.cn (K.Y.)
* Correspondence: 182030049@stu.just.edu.cn

Received: 26 July 2020; Accepted: 14 August 2020; Published: 17 August 2020

Abstract: As an effective carrier of renewable distributed power sources, such as wind power and photovoltaics, microgrids have attracted increasing attention as the energy crisis becomes more serious. This paper focuses on the symmetry between the dynamic optimal dispatch and the coordinated control of islanded microgrid to determine the optimal system configuration that can reliably meet energy needs. In order to solve energy management problems, operating costs and environmental benefits, a novel methodology that combines dynamic optimal dispatch and Grey Wolf Optimizer (GWO) is developed in this study to obtain the best output of different system components. This is to minimize the total cost of microgrid power generation and reduce pollutant emissions. In addition, a comparison is carried out between GWO and Particle Swarm Optimization (PSO). Moreover, the comparison between system configurations in six different scenarios and the effectiveness of GWO in solving optimization problems are presented. Finally, the simulation results show that GWO is more effective than PSO in determining the optimization parameters and the utilization rate of renewable energy in different scenarios is up to 92.96%. The simulations and experimental results verify the successful performance of the research method proposed in this study.

Keywords: dynamic optimal dispatch; wind turbine; photovoltaic; Grey Wolf Optimizer (GWO); energy management

---

1. Introduction

As the energy crisis becomes more serious, renewable distributed power sources, such as wind power and photovoltaics have gradually been developed, and microgrids are attracting further attention [1]. A microgrid is a small power generation and distribution system that integrates distributed power sources, energy storage devices, loads, and protection devices, and with the characteristics of flexible, reliable and safe power supply [2].

In order to use all kinds of energy reasonably and effectively, the microgrid energy scheduling meets certain constraints and load demands, and rationally dispatch energy and energy storage devices, which can effectively reduce operating costs and improve environmental benefits [3]. The energy dispatch of the microgrid is a key content in the related research problems of the microgrid. The factors considered in the dispatch model will affect the final dispatch result. Its purpose is to reasonably allocate the various loads under the premise of meeting the normal demand of all loads [4]. The output of the unit minimizes the total operating cost of the microgrid, thereby achieving the best economic benefits [5]. Dey et al. [6] studied the economic dispatch of a grid-connected renewable integrated microgrid system. Yuan et al. [7] proposed an energy management strategy based on hybrid prediction for the data interruption. Xin Li, et al. [8] considered that the microgrid environment/economic dispatch is a complex multi-objective optimization problem and reduced specific requirements for algorithm performance. Tiaan et al. [9] studied a multi-objective optimization model for multi-microgrid systems, which can not only minimize operating costs, but also reduce emissions.

Most of these papers use a static optimization scheduling model that aim to minimize the operation cost of the microgrid. Since the research on the economic optimal scheduling of the microgrid focuses on the operating economy after the system is built, the construction investment cost of the microgrid are not considered in most models, and the correlation between each time period is usually ignored. Each time period is independently optomised [10]. Compared with traditional power grids, the optimal dispatch of microgrids is more complicated, and traditional optimization methods cannot achieve optimal dispatch results. Various optimization methods based on artificial intelligence have the characteristics of fast convergence speed and not easy to fall into the local optimum. These optimization methods mainly include: Genetic Algorithm (GA) [11,12], PSO [13], and GWO [14]. However, GA and PSO also have their own disadvantages. The non-directional mutation of GA is its basic disadvantage. Likewise the convergence speed of PSO is not fast enough. In addition, the diversity of PSO is not enough, and it takes a long time to adjust the parameters in the optimization strategy. This paper uses PSO and GWO to compare and verify the effectiveness and accuracy of GWO in optimal scheduling. Naderi et al. [15] used fuzzy based hybrid PSO-DE to perform multi-objective economic emission dispatch on 10, 40 and 160 unit systems consider power loss, ramp rate, prohibited operating zones and valve point effects. In [16], a new multi-objective GWO for Optimal Reactive Power Dispatch (MORPD) is studied, which minimizes voltage deviation and active power loss (http://hainan.weather.com.cn/skjc/index.shtml).

Although a lot of work has been done in the above research on optimal dispatch of microgrid systems, none has studied dynamic optimal dispatch in conjunction with GWO. Instead, they use other intelligent algorithms to solve the optimization problem or study the static optimal dispatch of the microgrid. This paper proposes a new technology that combines dynamic optimal dispatch and GWO to achieve symmetry between the lowest operating cost of microgrid system and coordinated control of various devices. In addition, the proposed GWO is compared with the classic PSO to prove the effectiveness of the proposed method for dynamic optimal scheduling of microgrid systems. The comparison clearly shows that GWO has better performance, and has very fast convergence and balance in system optimization, which can avoid local optimization. The case study in this paper takes the Sanya region of China as an example. The region has 2534 h of sunshine throughout the year and has sufficient light.

The rest of this paper is organized as follows: The mathematical modeling of the microgrid components is elaborated in Section 2. The establishment of the objective function is described in Section 3. The constraints in the system are described in Section 4. In Section 5, the strategy of dynamic optimization scheduling for the system is outlined, and the GWO is introduced. Finally, the simulation results of the case study are presented and analyzed.

2. Mathematical Modeling

The basic structure of the microgrid is shown in Figure 1. The structure of the microgrid system includes wind power generation system (WT), photovoltaic power generation system (PV), energy storage system (ess), diesel generator (DG), and four power loads. Among them, as the micro power source of the main output unit in the microgrid, wind and solar power generation units use natural energy to generate electricity [16], which does not produce any pollutants but as natural resources, they also have their own limitations [17]. Restricted by natural conditions, randomness, volatility and intermittent characteristics have also become the inherent attributes and difficulties of clean energy power generation [18]. As an important coordination part of the system, the battery energy storage system is mainly used to coordinate the supply and demand balance of the microgrid, which plays a role in cutting peaks and filling valleys and smoothing fluctuations [19,20]. The backup power supply is used as a supplement to the microgrid power generation unit to ensure that some important loads in the microgrid system can be continuously and uninterrupted in the event of an emergency. Finally, each distributed unit in the microgrid is controlled by the microgrid energy management system to coordinate to form a unified system and maintain safe and stable operation.

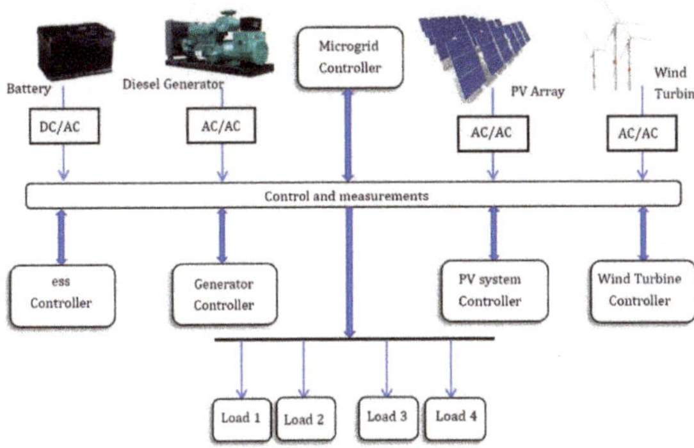

Figure 1. Basic structure diagram of islanded microgrid.

2.1. Model of Wind Turbine

Wind power generation is one of the important power generation units in the microgrid. As the objective of this paper is to optimize the microgrid, it is also necessary to predict the active power output by the wind turbine. The rotation speed of the hub is related to the wind speed, so the analysis of modeling the output power of the wind turbine is essentially an accurate measurement of the wind speed of the wind turbine hub [21]. Considering the measurement of the wind speed of the fan hub, the cut-in wind speed V_{ci}, rated wind speed V_r, and cut-out wind speed V_{co} are often used. Three physical quantities are measured, and then the fan output power characteristic equation is obtained by curve fitting. All wind turbines have roughly the same wind speed power curve shape. Total extracted power from the wind turbines P_{wt} at any time can be calculated as follows [22]:

$$P_{wt}(v) = \begin{cases} 0, & 0 \leq v \leq v_{ci} \\ a{\cdot}v^3 - b{\cdot}P_{wt-rate}, & v_{ci} \leq v \leq v_r \\ P_{wt-rate}, & v_r \leq v \leq v_{co} \end{cases} \tag{1}$$

The constants a and b are given by the following equations [22]:

$$a = \frac{P_{wt-rate}}{v_r^3 - v_{ci}^3} \qquad b = \frac{v_{ci}}{v_r^3 - v_{ci}^3} \tag{2}$$

where a and b respectively represent the fitting coefficients of the WT output power, and $P_{wt-rate}$ is the rated power of the WT. Generally, in the standard test case, the wind speed power characteristic curve of the wind turbine is drawn, and then the wind speed power expression shown in the above formula is obtained by curve fitting, but there are certain errors in the actual environment, so the standard test environment is correct the wind speed power characteristic curve obtained below.

2.2. Model of PV Array

PV power generation is one of the main power generation units in the microgrid system. In solving the problem of microgrid optimization and dispatch, it is necessary to accurately predict the power of photovoltaic power generation. Therefore, the output power of photovoltaic should consider solar radiation and temperature, the function is as follows [23]:

$$P_{pv}(t) = N_{pv}P_{rate-pv} * \frac{S}{S_{ref}} * [1 + K_t(T_c - T_{ref})] \tag{3}$$

where P_{pv} is the output power of PV arrays, N_{pv} is numbers of PV arrays, the maximum output power of the photovoltaic array is expressed in $P_{rate-pv}$. This value is the rated output power obtained by measuring the output of the PV array in a standard environment with a solar radiation intensity (S_{ref}) of 1 kW/m² and a temperature (T_{ref}) of 25 °C under no wind conditions. S is solar radiation intensity and T_c is the PV cell temperature.

The PV cell temperature can be calculated as follows [23],

$$T_c = T_a + S(\frac{NOCT - 20}{800}) \tag{4}$$

where T_a is the ambient temperature and $NOCT$ is the temperature of the battery under standard operation.

2.3. Model of Energy Storage System (ess)

The ess can store electric energy when the electric energy is sufficient, and release electric energy when the electric energy is insufficient. The state of charge of the battery is divided into two types: charging and discharging, which increases the flexibility and reliability of the microgrid. The power of the battery is as follows [23,24]:

$$E_b(t) = E_b(t-1) + [P_{pv}(t) + P_w(t) + P_{ch,t}]\eta_b^{ch} \tag{5}$$

where $E_b(t)$ and $E_b(t-1)$ are the power stored in the battery at times t and $t-1$, $P_{ch,t}$ represents the battery charging power, η_b^{ch} represents the battery charging efficiency, generally take 95%.

Besides, when the load demand is large, the power of the system cannot meet the load demand, the battery is in a discharged state. Therefore, the energy of the battery at the time t can be expressed as follows [24],

$$E_b(t) = E_b(t-1) - [P_{pv}(t) + P_w(t) + P_{dch,t}]/\eta_b^{dch} \tag{6}$$

where η_b^{dch} represents the battery discharge efficiency, in this study, it is taken 100%. $P_{dch,t}$ represents the battery discharging power.

For the modeling of the above mentioned, there are still many constraints, such as the mutual repulsion constraint of the battery's charge and discharge state, the constraint of the state of charge, and the constraint of charge and discharge power are as follow [23,24]:

$$\begin{aligned} E_{b\ min} &\leq E_b(t) \leq E_{b\ max} \\ E_{b\ min} &\leq (1 - DOD)E_{b\ max} \\ E_{b\ max} &= N_{batt}E_{rate-batt} \end{aligned} \tag{7}$$

where N_{batt} is the number of battery, E_{bma} and E_{bmin} are the maximum and minimum storage capacity, and $E_{rate-batt}$ is the battery pack rate (kWh), and DOD is the depth of discharge, which is taken 80% in this study.

2.4. Model of Diesel Engine

Considering the unpredictability and uncontrollability of wind turbines and photovoltaic power generation in the microgrid system, in order to meet the reliability of the power supply of the microgrid system, it is usually necessary to configure a backup power supply for the microgrid system in case of emergency [25]. The backup power source configured in this study is a diesel generator.

However, diesel generators will cause environmental pollution and increase the operating cost of the system. Therefore, it is usually only put into use when renewable energy generation is insufficient to meet the power demand of the load. Moreover, diesel generators cannot run at lower operating power levels. Operating power levels that are too low not only increase fuel consumption, but also affect the operation of diesel generators, and reduce their service life. The minimum operating power level of the generator is 30% in this paper.

In addition, diesel generators should be able to operate at a power level of 75%. At this time, it is not only the most economical in terms of fuel consumption and output power, but also the spinning reserve of the unit, which is the best operating power for diesel generators.

In summary, in the range of 30–100% of diesel generator operating power, the relationship between its fuel consumption F and its output power can be expressed as follows [25],

$$F = F_0 \cdot P_{de-rate} + F_1 \cdot P_{de} \tag{8}$$

where $P_{de-rate}$ and P_{de} represent the rated power value and actual output power value of the diesel engine, respectively, F_0 and F_1 represent the two fitting coefficients of the fuel-power curve of the diesel generator, which can generally be measured according to the actual measurement of the diesel generator.

3. Objective Function Formulation

This paper considers the overall system operation cost as the objective for the microgrid optimization. It mainly considers the operation and maintenance costs of wind turbines, photovoltaics, diesel generators and energy storage systems, the depreciation costs and energy loss costs of battery energy storage systems, and the emission costs brought about by the operation of diesel generators.

3.1. Cost Analysis of Distributed Power

Because wind and solar are clean energy, regardless of the cost of power generation, so wind turbines and photovoltaic power generation systems mainly consider equipment maintenance costs [26]. Diesel generator operation needs to consider its power generation costs and operation and maintenance costs. Wind turbine and photovoltaic equipment maintenance costs are as follow [26],

$$\begin{aligned} C_{om-pv} &= c_{m-pv} \cdot P_{pv,t} \\ C_{om-wv} &= c_{m-wt} \cdot P_{wt,t} \\ C_{de} &= (a \cdot P_{de,t}^2 + b \cdot P_{de,t} + c) + c_{om-de} \cdot P_{de,t} \end{aligned} \tag{9}$$

where c_{m-pv}, c_{m-wt} and c_{om-de} represent the unit power maintenance costs of photovoltaic power generation units, wind turbines and diesel generators, respectively, c_{om-de}, $P_{wt,t}$ and $P_{de,t}$ represent photovoltaic power generation units, the rated power output of the generator and the diesel generator at time t, a, b, and c respectively represent the power generation fitting coefficients of the diesel generators.

3.2. Analysis of Operating Cost of Energy Storage System

The operation and maintenance costs of the energy storage system can be divided into fixed parts and variable parts. The former is related to the rated capacity of the energy storage system, and the latter is related to the cumulative power generation of the energy storage system, which can be calculated as follows [26,27],

$$C_{om-ess} = c_{m-ess} \cdot R_{r-ess} + c_{me-ess} \cdot E_a \tag{10}$$

where $c_{m-ess} \cdot R_{r-ess}$ represents a fixed part of the operation and maintenance cost of the energy storage system, c_{m-ess} and R_{r-ess} represent the unit operation and maintenance cost and rated capacity of the ess, $c_{me-ess} \cdot E_a$ represents the variable operation and maintenance of the energy storage system cost [27]. In addition to the operation and maintenance costs of the ess, the depreciation cost C_{b-ess} of the ess and the power loss C_{lo-ess} are also considered.

3.3. Analysis of Emissions

The diesel generator consumes fuel to generate polluting gas during the power generation operation, such as CO_2, SO_2, NO_X, etc.

In order to count and reduce the emissions of these gases, it is necessary to reduce the start-up operation of diesel generators, according to the different degrees of different types of pollution to the atmosphere, so the unit gas emission treatment costs are set, similar to the form of a penalty function to generate environmental protection costs, expression as follows [26],

$$C_{pol} = \sum_n \varphi_n \cdot V_n = \sum_n \varphi_n \cdot V'_n \cdot P_{de} \tag{11}$$

where C_{pol} represents environmental protection costs; n indicates the type of harmful gas emitted, such as CO_2, SO_2, etc.; φ_n represents the unit treatment cost of a certain harmful gas; V_n indicates harmful gas n emissions; V'_n means diesel generator exhaust gas per unit power. For the purpose of calculation, the emissions from the operation of diesel generators are linearly closed.

After analysis, taking into account that diesel generators have many influencing factors on pollutants emitted during operation, including diesel generators, operating conditions, ambient temperature, and quality of diesel, etc. [28]. In order to facilitate calculation, this article sets the main harmful gas of diesel generators as CO_2, SO_2 and nitrogen oxides. For detailed parameter settings, see the analysis of examples in this paper. In summary, the objective function established in this part is:

$$minC = min[C_{om-pv} + C_{om-wv} + C_{de} + C_{om-ess} + C_{pol}] \tag{12}$$

4. Constraints

For the above objective function. The main constraints established in this section are as follow:

(1) Power balance constraint,

$$P_{pv,t} + P_{wt,t} + B_{dis,t} \cdot P_{dis,t} + P_{de,t} = P_{l,t} + B_{ch,t} \cdot P_{ch,t} \tag{13}$$

where $P_{pv,t}$ and $P_{wt,t}$ represent the output power of the WT and photovoltaic at time t, $P_{l,t}$ represent the load power at t, and $P_{de,t}$ represent the output power of the diesel generator at t.

(2) The output power constraint of the diesel generator,

$$P_{de-min} \leq P_{de,t} \leq P_{de-max} \tag{14}$$

(3) Battery energy storage constraints.

Energy storage system charge and discharge power constraints:

$$\begin{aligned} 0 \leq P_{ch,t} \leq P_{ch,max} \\ 0 \leq P_{dis,t} \leq P_{dis,max} \end{aligned} \tag{15}$$

Energy storage system charge state constraints:

$$S_{ess,min} \leq S_{ess,t} \leq S_{ess,max} \tag{16}$$

Mutually exclusive constraints of energy storage systems:

$$0 \leq B_{ch,t} + B_{dis,t} \leq 1 \tag{17}$$

5. Formulation of the Optimization Strategy

In order to solve the problem of optimal operation and scheduling of islanded microgrid, it is usually more effective to use the energy optimization management method with multi-period coordination. The flow chart in Figure 2 demonstrates the scheduling strategy proposed in this paper. Since the islanded microgrid system can only use the power output power of the WT and PV power generation system, it is necessary to predict the wind, solar and load demand in the future in advance. Due to the volatility of wind and solar energy, further short-term forecasting of wind and solar energy is needed to ensure the accuracy of the forecast. After obtaining the forecast data of wind energy, solar energy and load, it is divided into 6 different scenarios.

Scenario 1: When the electrical energy generated by the WT and PV can meet the demand of the load, it should then be determined whether the ess needs to be charged. When the state of charge is sufficient and charging is not required, the power output of the WT and PV is limited by abandoning the wind and the light.

Scenario 2: When the state of charge of the ess is insufficient and charging is required, it is further determined whether there is excess electrical energy for energy storage. If there is excess electrical energy, the ess is charged after meeting the load demand.

Scenario 3: If there is no excess power, the WT and PV output only need to meet the power supply of the load.

Scenario 4: If there is no excess electrical energy, the WT and PV output can only meet the load power supply. When the power generated by the WT and PV is insufficient to meet the load demand, it is necessary to determine whether the ess can be discharged to supplement the power. If ess does not have enough power to power the load, then need to start diesel generators to power the system load.

Scenario 5: If ess can supply power, it needs to further determine whether the total output of WT/PV and ess meet the load demand. If the output meets the need, then the wind, solar and energy storage system is used to supply power to the load.

Scenario 6: If the output cannot meet the load, it also needs to start diesel generators to supply power to the system load.

The flow chart in Figure 2 is the scheduling strategy proposed in this section.

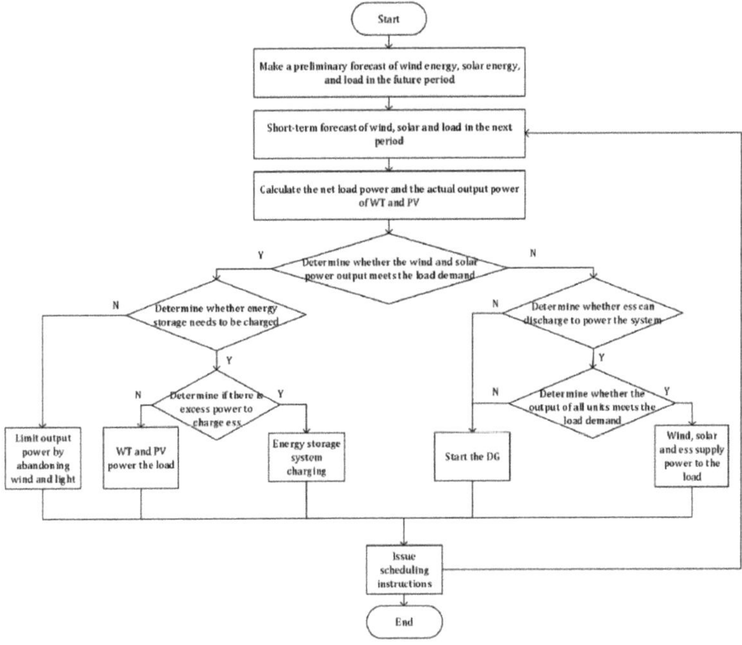

Figure 2. The scheduling strategy of the microgrid.

Optimal Sizing of Microgrid Using GWO

Grey Wolf Optimization (GWO) is a group intelligence optimization algorithm proposed by Griffith University scholar Mirjalili and others in Australia in 2014. The algorithm is an optimized search method developed by the grey wolf predator activity. It has the characteristics of strong convergence performance, few parameters, and easy implementation [29]. Grey wolves belong to canines that live in groups and are at the top of the food chain. The grey wolf strictly observes a hierarchy of social dominance. As shown in Figure 3.

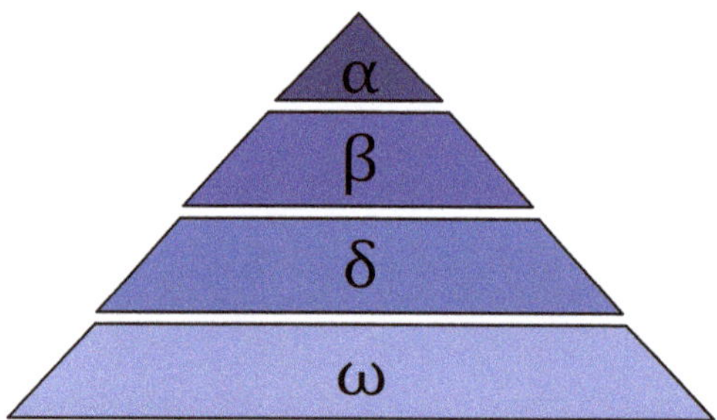

Figure 3. Grey wolf social dominance hierarchical relationship.

The GWO optimization process includes five steps. The specific steps are as follows.

(a) Social Hierarchy

First, wolf pack and set the number are initialized, then the fitness value of each individual in the wolf pack are calculated. Mark the grey wolves with the top three fitness values as α, β, δ, and the remaining wolves as ω. That is to say, the social rank in the grey wolf group is ranked from high to low in order of α, β, δ, and ω. The three optimal solutions (α, β, δ) in each iteration guide the optimization process of GWO.

(b) Encircling Prey

The grey wolf will gradually approach the prey and surround it when it searches for the prey. The functional expression for this behavior is as follows [29]:

$$
\begin{aligned}
D &= C \circ X_p(t) - X(t) \\
X(t+1) &= X_p(t) - A \circ D \\
A &= 2a \circ r_1 - a \\
C &= 2r_2
\end{aligned}
\tag{18}
$$

where t is the current number of iterations, A and C are the synergy coefficients; vector X_p represents the position vector of the prey; $X(t)$ represents the current grey wolf's position vector; a linearly decreases from 2 to 0 during the entire iteration process; r_1 and r_2 are the random vector in [0, 1].

(c) Hunting

The grey wolf has the ability to identify the position of the potential prey (optimal solution). However, the solution space characteristics of many problems are unknown, and the grey wolf cannot determine the precise position of the prey.

In order to get the best optimization plan, it is assumed that α, β, δ have the ability to identify the possible location of prey to simulate the behavior of grey wolf. Therefore, keep the best three grey wolves (α, β, δ) in the current population during iterating, then update their positions according to the positions of other search agents (including ω). The mathematical model of this behavior can be expressed as follows [29]:

$$D_\alpha = C_1 \circ X_\alpha - X, \ D_\beta = C_2 \circ X_\beta - X, \ D_\delta = C_3 \circ X_\delta - X$$
$$X_1 = X_\alpha - A_1 \circ D_\alpha, \ X_2 = X_\beta - A_2 \circ D_\beta, \ X_3 = X_\delta - A_3 \circ D_\delta \qquad (19)$$
$$X(t+1) = \frac{X_1 + X_2 + X_3}{3}$$

where X_α, X_β, X_δ represent the position vector of α, β, δ in the current population; X represent the position vector of the grey wolf; D_α, D_β, D_δ represent the distance between the current search agent and the best three wolves; when the $|A>1|$, the gray wolf searches for prey in different areas as much as possible. When $|A<1|$, grey wolves focused on searching for prey within a certain area.

(d) Attacking Prey

In the process of constructing the attacking prey model, according to (b), the decrease of a value will cause the value of to fluctuate accordingly. In other words, is a random vector in the interval $[-a, a]$. When is in $[-1, 1]$ interval, the search agent's position can be anywhere between the current grey wolf and its prey at the next moment.

(e) Search for Prey

Grey wolves mainly rely on the information of α, β, and δ to find their prey. In the process of searching for prey, keeping the search agent away from the prey can make the grey wolf perform a global search. In formula (b), the C vector composed of random values in the interval range $[0, 2]$. The random search behavior of grey wolves can make the optimization results more accurate and avoid falling into local optimum. C is a random value during the iteration process. This coefficient is helpful for the algorithm to jump out of the local area, especially the algorithm is particularly important in the later stage of the iteration.

After the microgrid obtains real-time information on the system status, it begins to execute the GWO. First, it sets the number of wolves and initializes the wolves, and then calculates the fitness value of each wolf. The top three are recorded as α, β, δ each wolf updates its position by calculating the distance from α, β, δ and finally outputs the global optimal solution, according to whether the maximum number of iterations is reached. Some of the previous optimization algorithms are prone to fall into the shortcomings of local optimization, slow convergence, and optimization of the microgrid. The flow chart in Figure 4 is the microgrid dispatching process combined with the GWO.

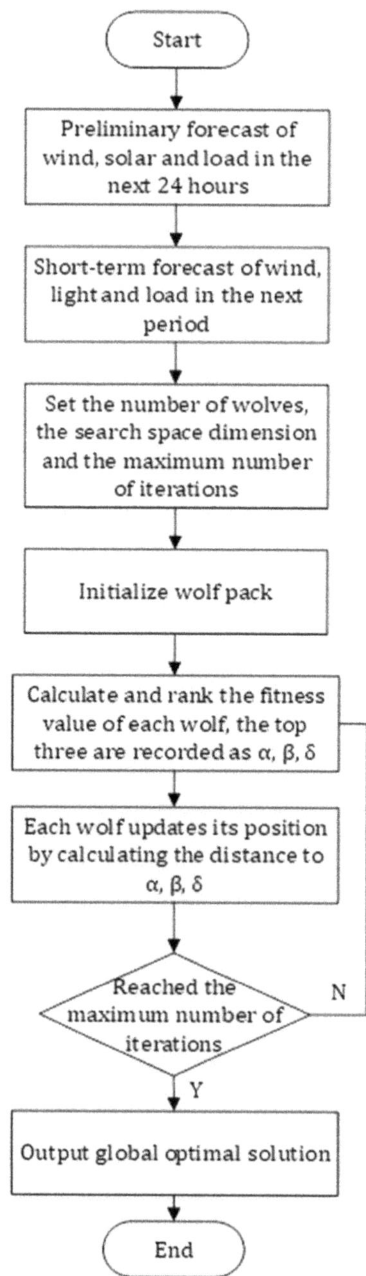

Figure 4. Flowchart for microgrid dispatching process combined with the GWO.

6. Case Study and Simulation Results

Sanya is located south of the Tropic of Cancer and has a tropical monsoon climate characterized by high temperature and rain. All relevant data are obtained from the official website of the local meteorological bureau. The annual average temperature is 26.7 °C. The highest temperature month is June with an average of 29.7 °C. The lowest temperature month is January with an average of 22.4 °C. The sunshine time of the year is 2534 h. The average annual precipitation is 1347.5 mm. Known as the "natural greenhouse".

6.1. Case Study

Figures 5 and 6 show the hourly wind speed and solar radiation data. However, due to the instability of wind energy and the tendency of wind density to change, this brings certain challenges to research. In order to ensure that the energy generated by the system can be balanced with the load demand. In this paper, we use the Artificial Neural Networks(ANN) to forecast the wind power. Figure 7 shows the regression graph obtained using ANN training test data including wind speed and solar radiation. We see that 0.96316 in the 40th iteration, which demonstrates a high correlation between the results obtained after training and the target. In addition, most of the results generated by the training data are related to the best fit line.

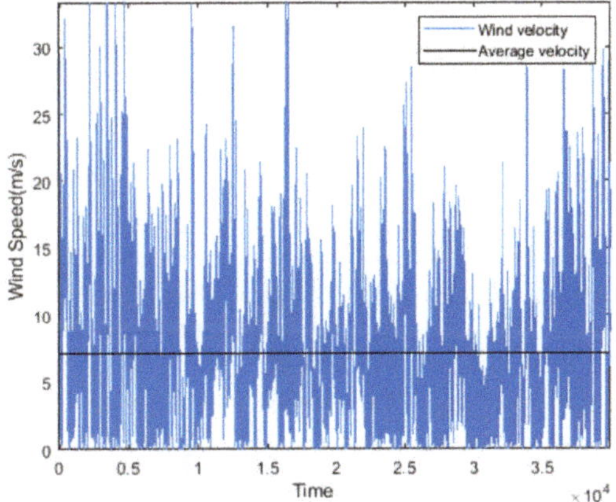

Figure 5. Annual wind speed of the studied location.

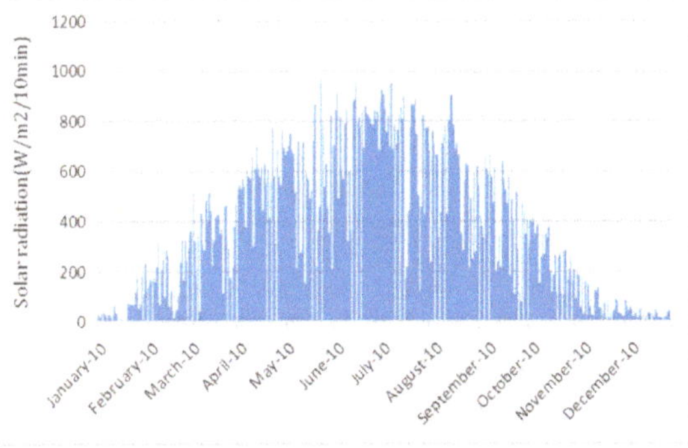

Figure 6. Annual solar irradiance of the location.

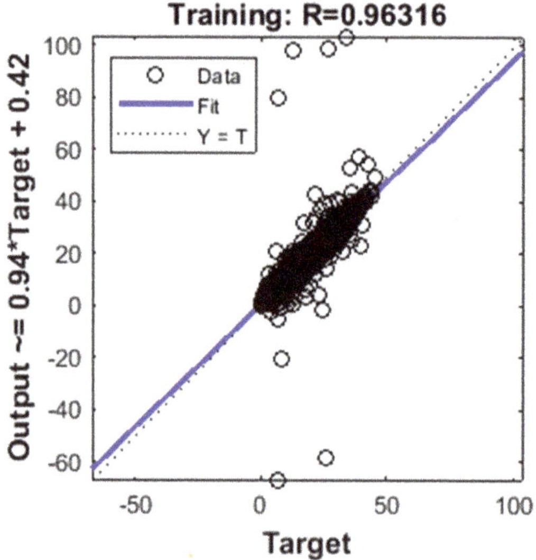

Figure 7. Regression of wind power ANN.

6.2. Description of the System

The basic parameters of the WT, PV and diesel generator equipment in the microgrid system are shown in Table 1. The rated capacity of the configured single battery is 50% of the total installed capacity of PV and WT [30]. Other parameters of the battery are shown in Table 2. In this paper, there are three main types of air pollutants emitted during the operation of diesel generators: CO_2, NOx and SO_2 [31]. The corresponding emissions and environmental treatment costs are shown in Table 3.

Table 1. Basic parameters of WT, PV and DG.

Components	Installed Capacity (kW)	UP Cost ($/kW)	UP Operation and Maintenance Cost ($/kW)	Power Generation Cost ($/kW)		
				a	b	c
WT	100	112	0.0042		0	
PV	150	24	0.0013		0	
DG	100	12	0.0245	0	0.24	0.0035

Table 2. Energy Storage System Parameters.

Parameter	Value	Parameter	Value
type	ess	Max charge & discharge power/kW	50
Capacity/kWh	50×5	Initial capacity/kW	50
Max allowable state of charge	95%	Charging efficiency	15
Min allowable state of charge	15%	Discharging efficiency	95%

The maximum discharge depth is 80%, and the operation and maintenance cost coefficient is 0.009 $/kWh [32].

Table 3. Corresponding Environmental Value of Pollutants.

Pollution	CO_2	NOx	SO_2
Emissions (g/kWh)	649.05	9.33	0.46
Value ($/KG)	0.0288	8.9747	2.1328

6.3. Simulation Results

In this paper, MATLAB R2019a is used for simulation and comparison of results. The response comparison of the microgrid system under different algorithms is shown in Figures 8 and 9. It can be seen from Figure 8 that using the GWO algorithm to solve the optimization scheduling problem of the microgrid is faster than the standard PSO algorithm, and it is easier to obtain the optimization results. Figure 9 shows the comparison of the parameter space of the two algorithms. It can be seen from Figures 8 and 9 that GWO has a faster convergence speed, and it is easier to obtain optimization results, and there is no local optimal situation in the figure. Table 4 shows the standard and average solutions of GWO are better than PSO, and compared with the PSO algorithm, GWO does not show the worst solution with a large deviation from the optimal solution. Its standard deviation is also much smaller than the standard deviation of PSO. As seen in Figure 8, GWO convergence speed is faster than PSO, it is not easy to fall into the local optimal, obtaining the global optimal solution is faster, the result is better, and the efficiency of the algorithm is higher.

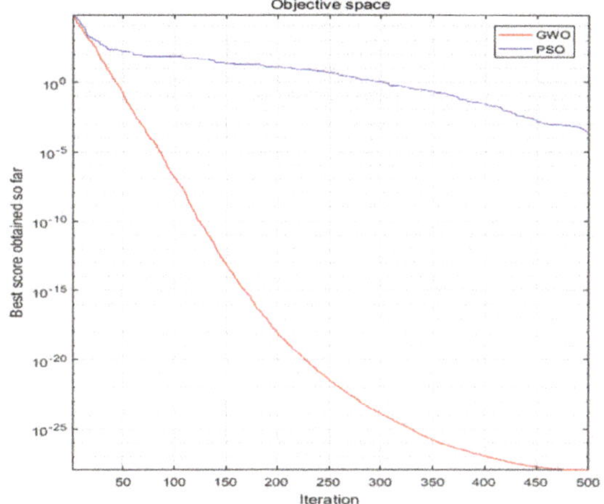

Figure 8. A comparison between algorithms.

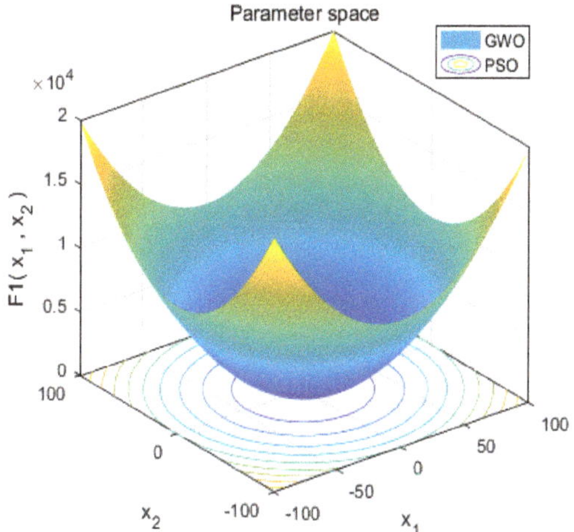

Figure 9. The parameter space of the two algorithms.

Table 4. Simulation Results of Test System Error.

	GWO		PSO	
	Ave	Std	Ave	Std
S1	−10.086523	12.233906	0.000136	0.000202
S2	2.517264	0.029014	0.042144	0.045421
S3	2.943147	79.14958	70.12562	22.11924
S4	8.249561	1.315088	1.086481	0.317039
S5	0.816579	0.000126	0.000102	17.50737
S6	0.002213	0.100286	0.122854	0.044957

The simulation results of the algorithm are better than PSO, which verifies that this paper is effective in optimizing the scheduling of microgrid system with GWO, and has superiority in convergence speed and optimization results compared with PSO.

6.4. Analysis of Optimal Dispatching Results of Microgrid

In Section 5, the optimized scheduling strategy proposed in this paper is presented, and the optimized scheduling strategy of microgrid with multiple time periods is used to divide the optimized scheduling strategy into 6 scenarios. In this part of the study, the output of each group of equipment in these 6 scenarios will be demonstrated, and the costs and pollutant emissions in each scenario will be compared.

Scenario 1 will determine whether the battery energy storage system needs to be charged when the power generated by WT and PV can meet the load demand. When the battery is in a sufficient state of charge and does not need to be charged, the power output of WT and PV is limited by abandoning wind and light. The predicted and the actual wind and solar values at each moment in Scenario 1 are shown in Figure 10. Under the condition of the Load1 which indicates the normal load size, the best output under constraints in Scenario 1 is shown in Figure 11. It can be seen from the figures that when the wind and solar resources are sufficient, while the ess system does not need to be charged, and the load demand is not large, the output of WT and PV can meet the load demand, so it does not need to run the diesel generator.

In Scenario 2, it is carried out under the same normal load demand as in Scenario 1 while the battery has insufficient power and needs to be charged. It is necessary to further determine whether the wind and solar power generation has excess power after supplying the load to charge the battery, Scenario 2 shows that there is excess electric energy to charge the ess, so the ess is charged after the load demand is met. The predicted wind and solar values are the same as shown in Figure 10. The best output of WT and PV under constraints in Scenario 2 is shown in Figure 12.

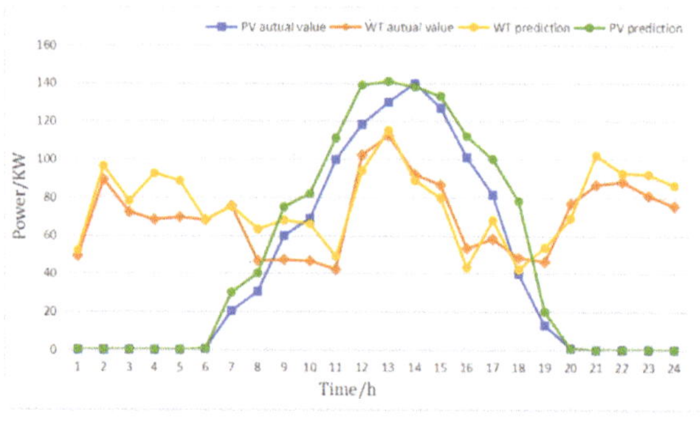

Figure 10. The predicted and actual value of WT and PV in Scenario 1.

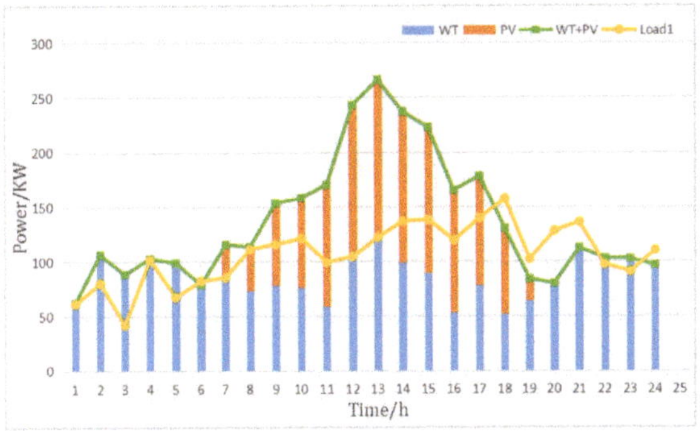

Figure 11. The best output under constraints in Scenario 1.

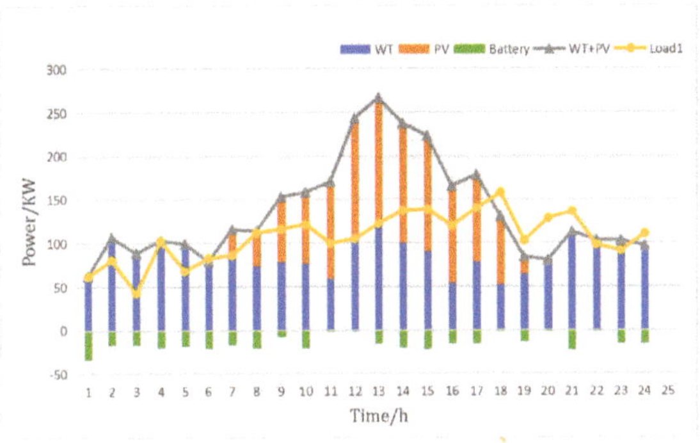

Figure 12. The best output under constraints in Scenario 2.

It can be found in Figure 12 that part of the wind and solar power generation capacity can charge the ess when the load demand is not large, which can consume more clean energy and the ess plays a role in the coordinated control of the system.

Scenario 3 shows that when the load demand is not large and the energy storage system needs to be charged, but if there is no excess electric energy to charge the energy storage system, the output of WT and PV only needs to meet the power supply demand of the load. The predicted and the actual wind and solar values at each moment in Scenario 3 are shown in Figure 13. Under the condition the best output under constraints is shown in Figure 14.

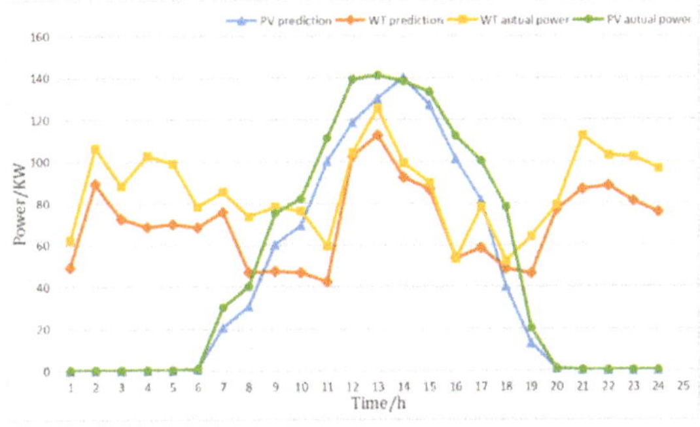

Figure 13. The predicted and actual value of WT and PV in Scenario 3.

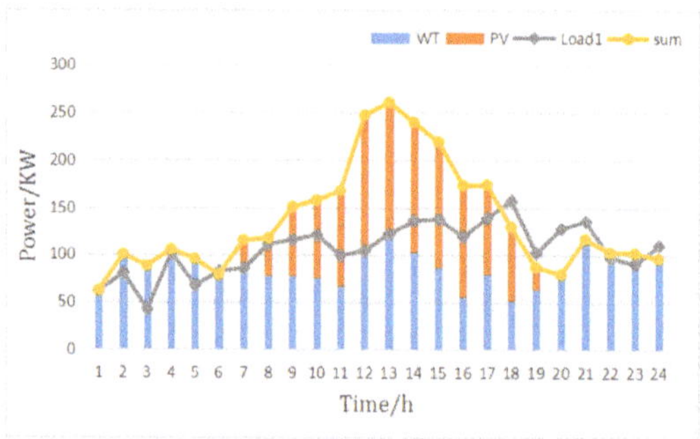

Figure 14. The best output under constraints in Scenario 3.

Different from the first three scenarios, scenarios 4–6 are carried out when the load demand is large. In Scenario 4, the wind and solar power generation is insufficient to meet the load demand, so it is necessary to determine whether the ess can discharge to give the load power supply, when ess does not have enough power to power the load, then need to start the diesel generator to power the load. Figure 15 shows the predicted and the actual wind, solar and DG power at each moment in Scenario 4. Figure 16 shows the best output under constraints.

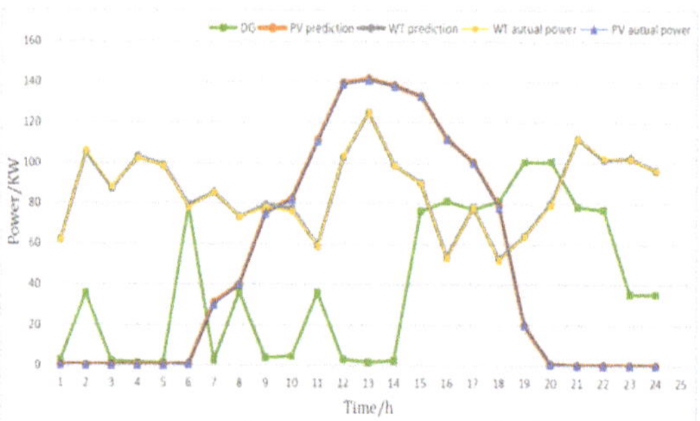

Figure 15. The predicted and actual value of WT, PV and DG in Scenario 4.

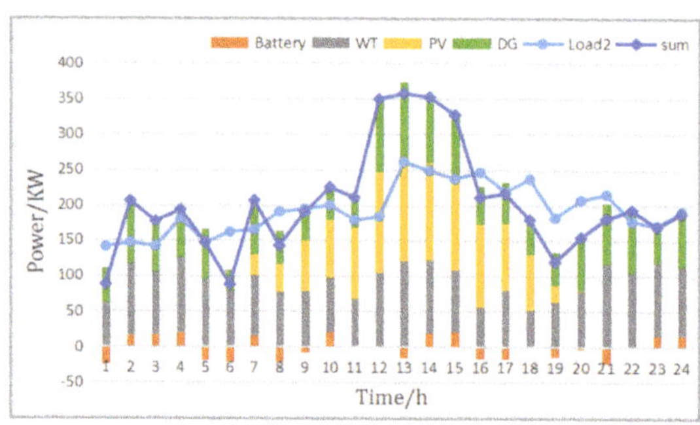

Figure 16. The best output under constraints in Scenario 4.

In Scenario 5, the load demand is large but ess has energy storage to supply power to the load and WT, PV and ess can meet the load demand. In this scenario, the predicted output value and actual output power of the scenery are shown in Figure 17, the best output under the conditions is shown in Figure 18.

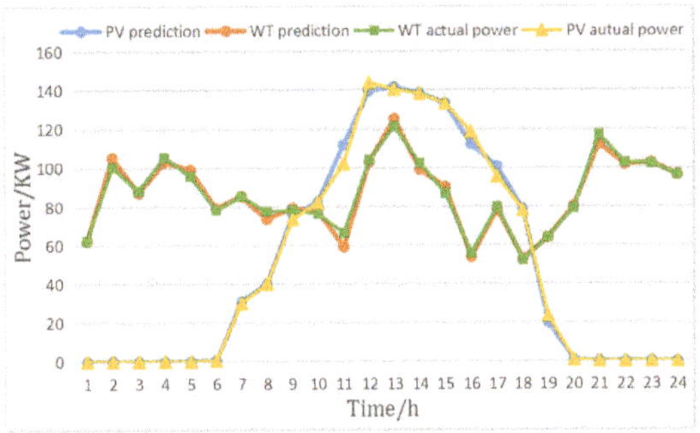

Figure 17. The predicted and actual value of WT and PV in Scenario 5.

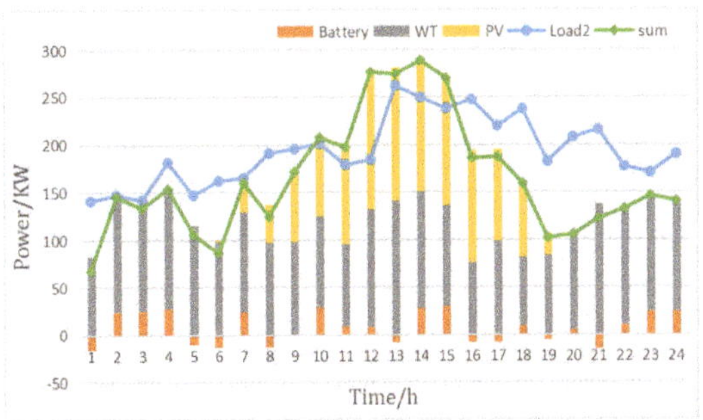

Figure 18. The best output under constraints in Scenario 5.

Scenario 6 is the last scenario of the study, and it is carried out under the condition of large load demand like Scenarios 4, 5, but at this time, ess has no electrical energy to power the load, so it is necessary to start the diesel generator to meet the load demand. The predicted wind and solar values are the same as shown in Figure 15. The best output of WT and PV under constraints in Scenario 6 is shown in Figure 19.

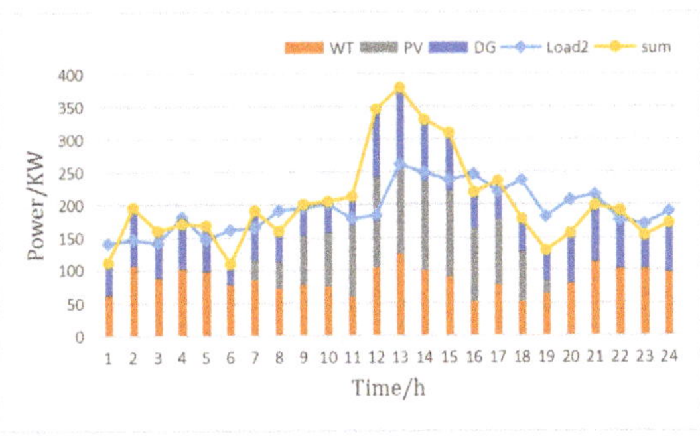

Figure 19. The best output under constraints in Scenario 6.

Under Load2, the wind power, PV power generation and energy storage systems can no longer meet the system load demand. The diesel generator in the system must be started, in order to make up for the power shortage of the system. The operating conditions are closely related, that is, there is a minimum output power and an optimal output power. In addition, the charge and discharge state of the ess under Load 2 is switched more frequently. From Figure 15, it can be seen that the diesel generator reaches full-running state from 19 to 20 o'clock. Compared with Load1, the consumption of wind power and photovoltaic power generation is higher under Load2 conditions, so the total cost is higher than the total cost under Load1. However, as Load2 requires more energy, making wind power and photovoltaic power generation so the energy utilization rate is also higher.

The comparison of the optimized scheduling results in 6 different scenarios is shown in Figure 20 and Table 5. As can be seen in Figure 20, when the load demand is larger, the utilization rate of clean energy is also higher, but from Table 5 it can be seen that the larger the load demand, the higher the total cost of the microgrid system, and the more environmental pollution emissions caused by starting the diesel generator. Among these 6 different scenarios, Scenarios 1, 3, and 5 are more ideal scenarios. In our daily life, diesel generators are often used to supply power to the load, so the emission of polluting gases is inevitable. The environmental cost of the system will increase accordingly. The comparison of optimized scheduling results in 6 different scenarios is shown in Figure 20.

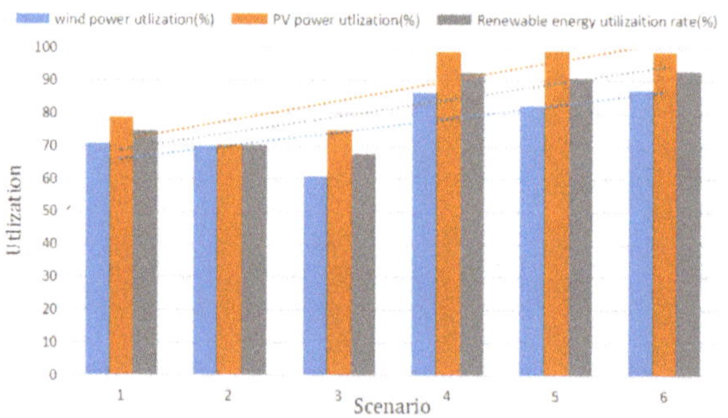

Figure 20. The comparison of optimized scheduling results in 6 different scenarios.

Table 5. The comparison of the system optimal scheduling results under 6 different scenarios.

Dispatching Results	Scenario 1	Scenario 2	Scenario 3	Scenario 4	Scenario 5	Scenario 6
Total power (kWh)	122,998	123,072	123,524	124,207	124,524	124,632
Total emissions (kg)	61,755	62,157	64,625	71,070	70,251	81,439
Total cost ($)	50,441	50,577	63,077	64,961	64,736	65,046
wind power (kWh)	10,941	10,941	10,941	36,327	36,327	36,327
PV power (kWh)	1330	1330	1330	6549	6549	6549
wind power utilization (%)	70.73	70.12	60.66	86.40	82.33	87.15
PV power utilization (%)	78.63	70.26	74.68	98.76	99.16	98.76
Renewable energy utilizaition rate (%)	74.68	70.19	67.67	92.58	90.75	92.96
Renewable energy output fuluctuations	26,891.16	28,121.92	4225.39	9859.25	15,568.06	10,322.18
Residual power generation capacity of WT and PV (kWh)	3486.65	3664.71	4640.95	5021.68	6473.99	4749.23

Table 5 shows the comparison of the system optimal scheduling results under 6 different scenarios. As can be seen from the table, when the load demand is large, the system's total power generation is the largest, but the total cost is also the highest, because the diesel generator is started. Therefore, the emission of polluting gas is also the most, so the cost of environmental governance will increase. However, on the other hand, due to the increase in electrical energy required, the utilization rate of renewable energy has reached a maximum of 92.96%, reducing wind and light, and the battery energy storage system has also played a role in cutting the valley and filling the peak effect.

7. Conclusions

This paper presents an optimal sizing of an islanded microgrid, optimized by dynamic optimization with GWO. The considered microgrid is a small autonomous system that integrates a variety of clean energy distributed power generation systems, energy storage systems, backup power sources, and electrical loads. Due to the complexity of optimal scheduling, a multi-objective optimal scheduling method combining GWO and dynamic optimal scheduling is proposed. Moreover, a comparison between the optimization capabilities of GWO and PSO on the one hand, and the system is divided into six different scenarios for comparison on the other, in order to understand the best output that achieves the lowest total cost of microgrid system operation and the highest clean energy utilization rate. Meteorological data is used, which is measured by the local meteorological bureau, in addition to artificial neural networks used to predict wind and solar energy in the future for a residential area in Sanya, China. The results show that the advantage of combining the GWO solution in optimizing multi-objective problems lies in that it reduces the calculation time and obtains the best function value compared with the PSO method alone. Furthermore, the findings of the study illustrate the economic and environmental feasibility of starting diesel generators under heavy load demand. Using ess as storage can better play the role of peak and valley filling, thereby reducing the total cost of the system. By considering diesel generators as part of the hybrid power system, the utilization of renewable energy can be improved. However, the use of diesel generators will produce polluting gas emissions. To solve this problem, multi-objective optimization based on GWO is applied. The optimal scale of the hybrid system including PV/wind/ess/diesel generators is a total cost of $64961 and 71070 kg.

In the future, the multi-objective optimization problem of microgrid can be further studied. Secondly, this paper studies the dynamic optimal dispatching of islanded microgrid. The economics of grid-connected microgrid and the utilization rate of clean energy generation must be studied in depth. Finally, because we use GWO for optimization, we can make the method more competitive by adjusting the scheduling strategy.

Author Contributions: Y.W. and C.L. supervised the work, Y.W. conceived the idea. Y.W., C.L., K.Y. performed the research. All authors have read and agreed to the published version of the manuscript.

Funding: This research was funded by the National Natural Science Foundation of China, grant number 51307074. And The APC was funded by Jiangsu University of Science and Technology.

Acknowledgments: This work is supported by the National Natural Science Foundation of China (51307074), Jiangsu Provincial Natural Science Foundation of China (BK20130466), Jiangsu University of Science and Technology Graduate Education Teaching Reform Project (103080605).

Conflicts of Interest: The authors declare no conflict of interest.

Nomenclature

A, C	synergy coefficients
C_{om-pv}	photovoltaic maintenance cost
C_{om-wt}	wind turbine maintenance cost
C_{de}	diesel generator cost
C_{om-ess}	battery cost
C_{pol}	environmental governance cost
E_b	energy stored in the battery [kWh]
E_{bma}	maximum storage capacity [kWh]
E_{bmin}	minimum storage capacity [kWh]
$E_{rate-batt}$	battery bank rate [kWh]
F	Generator's fuel consumption [L]
F_0	fitting coefficients
F_1	fitting coefficients
N_{pv}	number of photovoltaics
P_{wt}	total power of the wind turbines [kW]
$P_{wt-rate}$	rated power of wind turbines [kW]

P_{pv}	output power of photovoltaics [kW]
$P_{rate-pv}$	rated power of photovoltaics [kW]
$P_{ch,t}$	battery charging power [kW]
$P_{dch,t}$	battery discharging power [kW]
r_1, r_2	random vectors [0, 1]
S_{ref}	maximal solar radiation [kW/m2]
S	solar radiation intensity [kW/m^2]
T_c	PV cell temperature [°C]
T_{ref}	reference temperature [°C]
T_a	ambient temperature [°C]

Greek Symbols

$\alpha, \beta, \delta, \omega$	wolfs in GWO algorithm
η_b^{ch}	battery charge efficiency [%]
η_b^{dch}	battery discharge efficiency [%]

Abbreviations

ANN	artificial neural network
DOD	depth of discharge
ess	energy storage systems
GA	genetic algorithm
GWO	grey wolf optimizer
PSO	particle swarm optimization
PV	photovoltaics
WT	Wind turbine
T_{ref}	reference temperature [°C]
T_a	ambient temperature [°C]

References

1. Lasseter, R.H.; Piagi, P. Microgrid: A conceptual solution. In Proceedings of the IEEE Power Electronics Specialists Conference, Aachen, Germany, 20–25 June 2004; pp. 4285–4291.
2. Hamad, A.A.; Azzouz, M.A.; El-Saadany, E.F. Multiagent supervisory control for power management in DC microgrids. *IEEE Trans. Smart Grid.* **2016**, *7*, 1057–1068. [CrossRef]
3. Dunham, H.; Cutler, D.; Mishra, S.; Li, X. Cost-optimal evaluation of centralized and distributed microgrid topologies considering voltage constraints. *Energy Sustain. Dev.* **2020**, *56*, 88–97. [CrossRef]
4. Dai, R.; Liao, H.; Shi, Y.Z.; Liao, Y.; Xia, H.B.; Chen, M.Y. Economic dispatch method of microgrid considering demand response under time-of-use electricity price. *J. Chongqing Univ.* **2019**, *122*, 36–52.
5. Mileta, Ž.; Goran, D. Fuzzy expert system for management of smart hybrid energy microgrid. *J. Renew. Sustain. Energy* **2019**, *11*, 034101.
6. Dey, B.; Bhattacharyya, B.; Sharma, S. Optimal Sizing of Distributed Energy Resources in a Microgrid System with Highly Penetrated Renewables. *Iran. J. Sci. Technol. Trans. Electr. Eng.* **2018**, *43*, 527–540. [CrossRef]
7. Yuan, D.; Lu, Z.; Zhang, J.; Li, X. A hybrid prediction-based microgrid energy management strategy considering demand-side response and data interruption. *Electr. Power Energy Syst.* **2019**, *260*, 139–153. [CrossRef]
8. Li, X.; Tang, R.; Lai, J. A Knowledge based Multi-objective Optimization Strategy for Microgrid Environmental or Economic Scheduling problems. *Energy Procedia* **2019**, *182*, 2942–2947. [CrossRef]
9. Gildenhuys, T.; Zhang, L.; Ye, X.; Xia, X. Optimization of the Operational Cost and Environmental Impact of a Multi-Microgrid System. *Energy Procedia* **2019**, *182*, 3827–3832. [CrossRef]
10. Kumar, D.; Verma, Y.P.; Khanna, R. Demand response-based dynamic dispatch of microgrid system in hybrid electricity market. *Int. J. Energy Sect. Manag.* **2019**, *13*, 318–340. [CrossRef]
11. Ismail, M.S.; Moghavvemi, M.; Mahlia, T.M.I. Genetic algorithm based optimization on modeling and design of hybrid renewable energy systems. *Energy Convers. Manag.* **2014**, *85*, 120–130. [CrossRef]

12. Ko, M.J.; Kim, Y.S.; Chung, M.H.; Jeon, H.C. Multi-Objective Optimization Design for a Hybrid Energy System Using the Genetic Algorithm. *Energies* **2015**, *8*, 2924–2949. [CrossRef]

13. García-Triviño, P.; Llorens-Iborra, F.; García-Vázquez, C.A.; Gil Mena, A.J.; Fernández-Ramírez, L.M.; Jurado, F. Long-term optimization based on PSO of a grid-connected renewable energy/battery/hydrogen hybrid system. *Int. J. Hydrog. Energy* **2014**, *39*, 10805–10816. [CrossRef]

14. Azim, H.; Davide, A.G.; Farshid, K.; Fabio, B. Renewable Energies Generation and Carbon Dioxide Emission Forecasting in Microgrids and National Grids using GRNN-GWO Methodology. *Energy Procedia* **2019**, *159*, 154–159.

15. Naderi, E.; Azizivahed, A.; Narimani, H.; Fathi, M.; Narimani, M.R. A comprehensive study of practical economic dispatch problems by a new hybrid evolutionary algorithm. *Appl. Soft Comput.* **2017**, *61*, 1186–1206. [CrossRef]

16. Nuaekaew, K.; Artrit, P.; Pholdee, N.; Bureerat, S. Optimal reactive power dispatch problem using a two-archive multi-objective grey wolf optimizer. *Expert Syst. Appl.* **2017**, *87*, 79–89. [CrossRef]

17. Dong, Y.; Cai, Z. Research on control strategy of isolated island microgrid based on synchronous inverter technology. In Proceedings of the 16th Shenyang Scientific Academic Conference, Shenyang, China, 9–30 October 2019.

18. Fan, S.; Xiong, H.W. Dynamic reactive power optimization model of power grid considering VSC and DFIG. *Power Syst. Prot. Control.* **2019**, *36*, 28–36.

19. Wu, C.; Lin, S.; Xia, C.J.; Guan, H. Distributed optimal dispatch of microgrid group based on model predictive control. *Power Syst. Technol.* **2019**, *111*, 30–39.

20. Wu, S.; Yu, J.; Tian, Z.; Zhang, L.Z.; Yang, W.; Li, Y.; Peng, L.; Wu, W. Unified power flow algorithm for AC/DC hybrid power grid based on extended node method. *Power Syst. Prot. Control.* **2018**, *70*, 22–30.

21. Fu, Y.; Sang, Y. The EU's REserviceS project and its enlightenment for China's wind power and photovoltaic participation in grid frequency regulation. *Power Syst. Technol.* **2019**, *10*, 1–6.

22. Shan, C.; Gao, C.S.; Long, L.Z.; Tian, L.H. Dynamic dispatch optimization of microgrid based on a QS-PSO algorithm. *J. Renew. Sustain. Energy* **2017**, *9*, 045505.

23. Abdelshafy, A.M.; Hassan, H.; Jurasz, J. Optimal design of a grid-connected desalination plant powered by renewable energy resources using a hybrid PSO–GWO approach. *Energy Convers. Manag.* **2018**, *173*, 331–347. [CrossRef]

24. Diaf, S.; Diaf, D.; Belhamel, M.; Haddadi, M.; Louche, A. A methodology for optimal sizing of autonomous hybrid PV/wind system. *Energy Policy* **2007**, *35*, 5708–5718. [CrossRef]

25. Azad, H.B.; Mekhilef, S.; Ganapathy, V.G.; Modiri-Delshad, M.; Mirtaheri, A. Optimization of micro-grid system using MOPSO. *Renew. Energy* **2014**, *71*, 295–306.

26. Yin, C.; Wu, H.; Locment, F.; Sechilariu, M. Energy management of DC microgrid based on photovoltaic combined with diesel generator and supercapacitor. *Energy Convers. Manag.* **2017**, *132*, 14–27. [CrossRef]

27. Bilal, B.O.; Nourou, D.; Sambou, V.; Ndiaye, P.A.; Ndongo, M. Multi-objective optimization of hybrid PV/wind/diesel/battery systems for decentralized application by minimizing the levelized cost of energy and the CO_2 emissions. *Int. J. Phys. Sci.* **2015**, *10*, 192–203.

28. Bukar, A.L.; Tan, C.; Lau, K.Y. Optimal sizing of an autonomous photovoltaic/wind/battery/diesel generator microgrid using grasshopper optimization algorithm. *Sol. Energy* **2019**, *188*, 685–696. [CrossRef]

29. Fescioglu-Unver, N.; Barlas, A.; Yilmaz, D.; Demli, U.O.; Bulgan, A.C.; Karaoglu, E.C.; Atasoy, T.; Ercin, O. Resource management optimization for a smart microgrid. *J. Renew. Sustain. Energy* **2019**, *11*, 065501. [CrossRef]

30. Zhang, X.; Wang, X. A review of grey wolf optimization algorithm. *Comput. Sci.* **2019**, *46*, 30–38.

31. Zhao, B.; Zhang, X.; Chen, J.; Wang, C.; Guo, L. Operation Optimization of Standalone Microgrids Considering Lifetime Characteristics of Battery Energy Storage System. *IEEE Trans. Sustain. Energy* **2013**, *4*, 934–943. [CrossRef]

32. Das, A.; Ni, Z. A Computationally Efficient Optimization Approach for Battery Systems in Islanded Microgrid. *IEEE Trans. Smart Grid* **2017**, *9*, 6489–6499. [CrossRef]

Article

Asymmetric Compensation of Reactive Power Using Thyristor-Controlled Reactors

Martynas Šapurov [1,2,*]**, Vytautas Bleizgys** [1,2]**, Algirdas Baskys** [1,2]**, Aldas Dervinis** [1]**,
Edvardas Bielskis** [1,3]**, Sarunas Paulikas** [2]**, Nerijus Paulauskas** [2] **and Vytautas Macaitis** [2]

[1] State Research Institute Center for Physical Sciences and Technology, Sauletekio av. 3,
LT-10257 Vilnius, Lithuania; vytautas.bleizgys@ftmc.lt (V.B.); algirdas.baskys@ftmc.lt (A.B.);
aldas.dervinis@ftmc.lt (A.D.); edvardas.bielskis@su.lt (E.B.)
[2] Faculty of Electronics, Vilnius Gediminas Technical University, Naugarduko st. 41, 03227 Vilnius, Lithuania;
sarunas.paulikas@vgtu.lt (S.P.); nerijus.paulauskas@vgtu.lt (N.P.); vytautas.macaitis@vgtu.lt (V.M.)
[3] Department of Functional Materials and Electronics, Siauliai University, P. Visinskio str. 38,
76352 Siauliai, Lithuania
* Correspondence: martynas.sapurov@ftmc.lt; Tel.: +370-684-70305

Received: 4 May 2020; Accepted: 25 May 2020; Published: 27 May 2020

Abstract: The thyristor-controlled reactor (TCR) compensator for smooth asymmetric compensation of reactive power in a low-voltage utility grid is proposed in this work. Two different topologies of compensator were investigated: topology based on a single-cored three-phase reactor and topology with separate reactors for every phase. The investigation of the proposed TCR compensator was performed experimentally using a developed experimental test bench for 12 kVAr total reactive power. The obtained results show that employment of separate reactors for every phase allows us to control the reactive power in every phase independently, and that the TCR compensator with three single-phase reactors is suitable for smooth and asymmetric compensation of reactive power in a low-voltage utility grid.

Keywords: reactive power; thyristor-controlled reactor; air-gaped reactor; low-voltage utility grid; asymmetric compensation of reactive power; smooth compensation of reactive power

---

1. Introduction

The number of small grid-connected photovoltaic and wind power plants is constantly growing. These plants supply energy to low-voltage lines of the utility. The quantity of energy supplied by these power plants depends on natural conditions, which often change. On the other hand, the electrical grid loads in low-voltage lines are not just three-phase but single-phase as well, and many of them use reactive power. This leads to the fact that one of the main present-day problems of the electrical grid is compensation of the reactive power in the low-voltage grid [1–18] and since the low-voltage three-phase lines are often loaded asymmetrically [14,19–21], the reactive power, which has to be compensated, is different in different phases.

At present, most parts of reactive power compensation systems in low-voltage lines of the utility are based on the electro-mechanical commutation technology of capacitor banks, which are split into steps connected in parallel to the utility grid [22,23]. However, capacitor banks have fixed discrete reactive power capacity, i.e., the reactive power produced by the capacitor bank cannot be changed smoothly. Therefore, it is impossible to fully compensate reactive power of the utility grid using capacitor banks [24].

Smooth reactive power compensation can be achieved by employing static synchronous compensators (STATCOMs) based on a voltage source inverter [24–30]. STATCOM devices are capable of compensating for both capacitance and inductance reactive power. The inverter of a

STATCOM device produces PWM voltage to the utility grid employing a low pass filter. Capacitance reactive power is supplied if the magnitude of the voltage provided by the inverter is higher than the voltage magnitude of the utility grid. In cases when the magnitude of the inverter voltage is lower than the magnitude of the utility grid voltage, inductive reactive power is consumed. The STATCOM compensator based on inverter has a fast response time and is capable of full reactive power compensation. The main disadvantages are its high price [19,20,31] and that STATCOM devices can provide just symmetric compensation of reactive power in all three phases of the grid [32].

Smooth reactive power compensation can be performed by employing the static VAR compensator (SVC), a shunt-connected variable reactance, which either generates or consumes reactive power. The static VAR compensator is a power electronic device based on thyristor-switched capacitors (TSCs) for discrete control of generated reactive power and thyristor-controlled reactors (TCRs) for smooth control of consumed reactive power [24–27]. A TCR consists of a reactor and a bidirectional thyristor connected in series. Consumed inductive reactive power is controlled by variation of the thyristor firing angle α, where $\alpha = 90$ corresponds to full reactive power and $\alpha = 180$ corresponds to zero reactive power [24–27,33–37]. The first step in this approach is to overcompensate the utility grid using the TSC (to make the utility-grid load slightly capacitive) and after that, by consuming the required amount of inductive reactive power by the TCR, the total compensation of reactive power is achieved.

Despite the fact that research in the field of SVCs has been carried out for many years, the topic is still relevant. This fact is evidenced by many new publications devoted to the theory and application of SVCs, e.g., [4,8,12,38–43]. One of the directions of recent research works in this field is the expansion of SVC application areas [8]. A new area of expansion could be the development and application of the SVC for smooth asymmetric compensation of reactive power in low-voltage grids as a cheaper alternative to the inverter-based STATCOM compensator. The novelty of such work can be proved by the following facts:

1. No one on the market offers the SVC, which is based on TSCs and TCRs, for smooth asymmetric compensation of reactive power in low-voltage grids.
2. There are few publications dedicated to the SVC for smooth compensation of reactive power in low-voltage grids [5–9,44]. However, all these publications are dedicated to symmetric compensation of reactive power in all three phases, and in most of them, just the simulation results are presented.
3. The TCR compensator, which is an essential part of the SVC that allows us to achieve smooth compensation, is developed only for the symmetric compensation of the reactive power and practically is employed just for high and medium voltage lines of the utility grid [4,24–27,35,39,41,45–47].

The novelty of this work is that the proposed TCR compensator is capable of compensating reactive power in a three-phase low-voltage grid utility asymmetrically and that the proper operation of the proposed compensator is proved experimentally, using developed experimental reactors and the test bench of the compensator.

2. The Topology and Operation of the TCR Compensator

The block diagram of the experimental test bench for the investigation of the developed TCR compensator for smooth asymmetric compensation of reactive power for a low-voltage utility grid is presented in Figure 1. It consists of a three-phase power supply ($|U| = 230$ V, $f = 50$ Hz); commutation switches SW1–SW3; zero crossing circuits for each phase; thyristor switches T1–T3 for each phase; control block. The test bench was designed for the experimental investigation of the TCR compensator operation with the three-phase single-cored reactor and with separate reactors for every phase. Therefore, the single-cored three-phase Y-connected reactor (L1) with the middle point connected to neutral, three-phase Y-connected separate phase reactors (L2–L4) with the middle point connected to neutral and switches SW2 and SW3 for commutation of reactors were included into the structure of the reactive power compensator (Figure 1). The power quality analyzer and oscilloscope were used for the measurement of the reactive and active power and waveforms of utility-grid voltage and current. The investigation was performed in low-voltage lines of the utility for symmetric and asymmetric phase load reactive power compensation. The Δ connection of coils as well as Y-connection with an unconnected midpoint were not used because they are not suitable for asymmetric compensation of the reactive power.

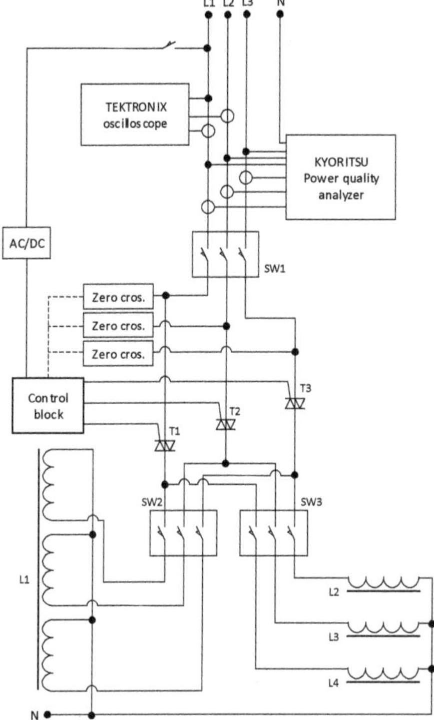

Figure 1. Block diagram of the thyristor-controlled reactor (TCR) compensator experimental test bench.

3. Investigation Results

The experimental investigation results of the proposed TCR compensator for smooth asymmetric compensation of reactive power in a low-voltage utility grid are presented in this section. Two different topologies of compensator are investigated: topology based on a single-cored three-phase reactor and topology with separate reactors for every phase. The main goal of investigations is to prove the possibility of smooth asymmetric compensation of consumed reactive power in a three-phase low-voltage utility grid using a TCR. The experimental test bench of the TCR compensator with a single-cored three-phase reactor is presented in Figure 2.

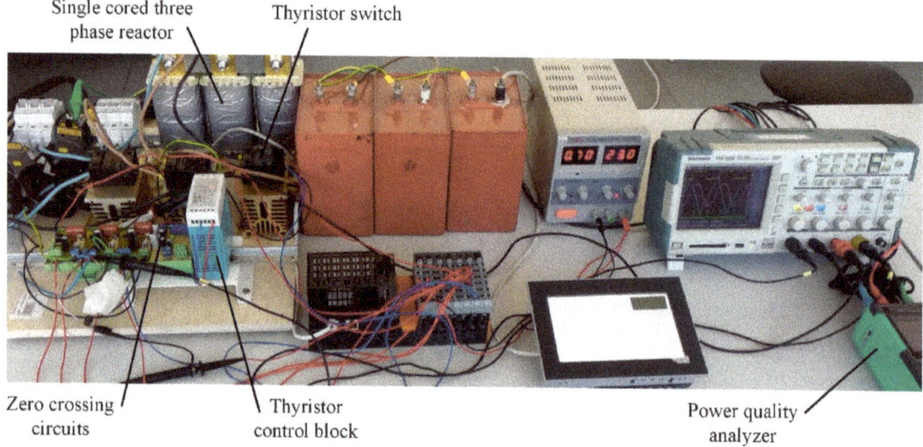

Figure 2. Experimental test bench of the TCR compensator with a single-cored three-phase reactor.

3.1. Investigation of the Compensator Based on a Single-Cored Three-Phase Reactor

The single-cored three-phase air-gaped reactor was designed in order to achieve sufficient reactive power compensation and low consumption of active power and to avoid core saturation. Cores with air gaps are usually used for reactors to avoid their saturation. The theory dedicated to the design of magnetic materials with the air gap, including the cores of reactors, can be found in [48–51]. The three-phase EI-shaped reactor was designed for total reactive power $Q = 4.8$ kVAr that corresponds to the RMS of phase current $I = 7A$ for the low-voltage utility-grid phase voltage $|U| = 230$ V. The impedance of the reactor has to be $|Z| = \frac{|U|}{|I|} \approx 32\ \Omega$. The inductive resistance of the coil was much higher than active; therefore, $Z \approx X_L$ and the approximate inductance of the coil can be obtained using the equation $L = \frac{X_L}{\omega} \approx 100\ mH$, where ω is the angular frequency of grid voltage. In order to avoid core saturation, the air-gaped core was used. To obtain the desired inductance of the reactor, the approximate design parameters of the reactor coil were chosen using the equation:

$$L = \mu_I\, \mu_0 \frac{N^2 \cdot S}{l},$$

(1)

where μ_0 is the free space permeability, μ_I is the relative magnetic permeability of iron core, N is the number of turns, S is the winding area and l is the length of coil. The parameters of the single-cored three-phase air-gaped reactor are presented in Table 1.

Table 1. Parameters of the single-cored three-phase air-gaped reactor.

Parameter	Value
Relative magnetic permeability of iron core (μ_I)	100
Number of turns of coil (N)	510
Winding area (S)	17.6 cm^2
Length of coil (l)	10.8 cm
Wire cross-section	1.8 mm^2
Inductance of coil at core air gap length $d = 0$	530 mH
Inductance of coil at $d = 6$ mm	100 mH
Inductance of coil at $d = 10$ mm	37 mH
Inductance of coil without core	5.3 mH

By adjusting the air gap, the inductance of every coil was set to 100 mH. The active resistance of every coil is $R = 1\ \Omega$. The active resistance in the equivalent circuit of the reactor was connected in series with the inductive one; therefore, the current was the same for both elements. The values of consumed active power $|P| = 54W$ and reactive power $|Q| = 1.7kVAr$ for each phase coil were determined by employing the following equations:

$$|P| = \frac{\left(|U| \cdot \frac{R}{|Z|}\right)^2}{R},$$ (2)

$$|Q| = \frac{\left(|U| \cdot \frac{X_L}{|Z|}\right)^2}{X_L}.$$ (3)

The structure and view of the designed single-cored air-gaped reactor are presented in Figure 3.

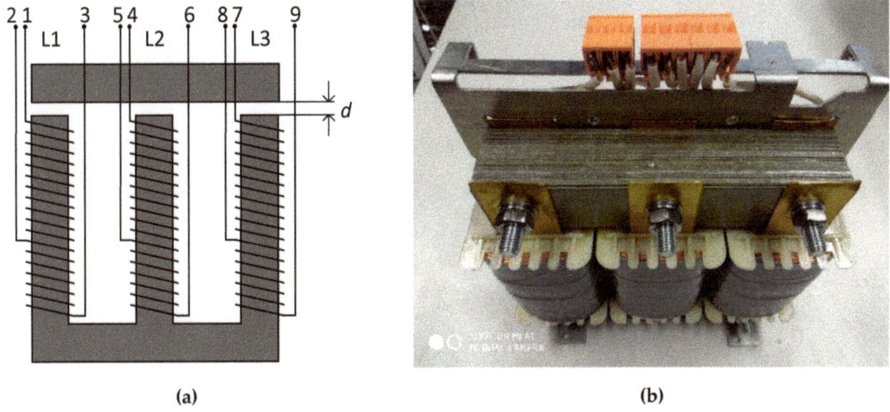

(a)   (b)

Figure 3. Structure (**a**) and view (**b**) of the single-cored three-phase air-gaped reactor.

The reactive power consumed by the reactor is determined by the duration the reactor is connected to the grid. This duration can be controlled by variation of the thyristor firing moment in relation to the grid-voltage zero-value-crossing moment. Usually this moment is named as the thyristor firing angle, α, and is expressed in angular degrees. The obtained reactive power dependences on the firing angle of thyristors, when firing angles in all three phases are changed simultaneously (in the case of symmetric compensation), are presented in Figure 4. The waveforms of utility-grid voltage, reactor current and thyristor firing pulses are given in Figure 5. The obtained results show that the reactive power consumed by the reactor changes in all phases by the same law. The dispersion of the reactive power between individual phases, which is about 17%, is caused by the dispersion of the parameters of the reactor coils. The obtained results allow us to conclude that the TCR compensator based on a single-cored three-phase reactor is suitable for the smooth symmetric compensation of reactive power in all three phases within appropriate error, determined by the dispersion of parameters of reactor coils.

The investigation results prove that the TCR technique can be implemented in a low-voltage utility grid for symmetric compensation of reactive power within appropriate error and that the reactive power consumed by the reactor can be controlled smoothly by control of the thyristor firing angle. Additionally, it can be stated that commutation of the reactor does not introduce any high-frequency disturbances of the reactor current and grid voltage (Figure 5) [52]. However, for the smooth control of the reactive power consumed by the reactor, it was necessary to pass the current only for a certain part of the period. As a result, the reactor current shape was distorted (Figure 5), resulting in low-frequency harmonics.

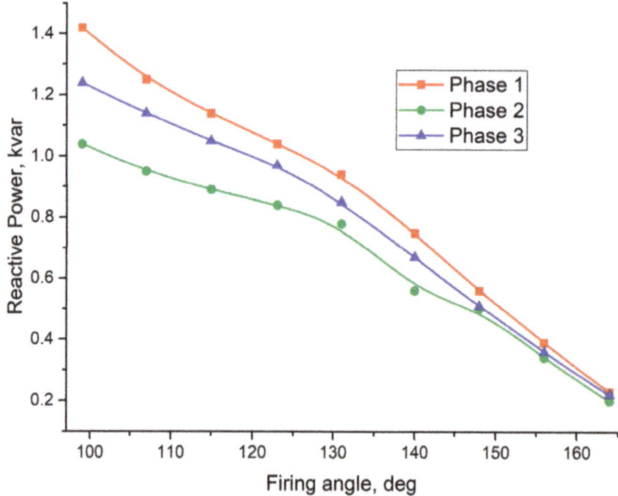

Figure 4. Dependences of reactive power consumed by the single-cored three-phase reactor on the firing angle of the thyristors.

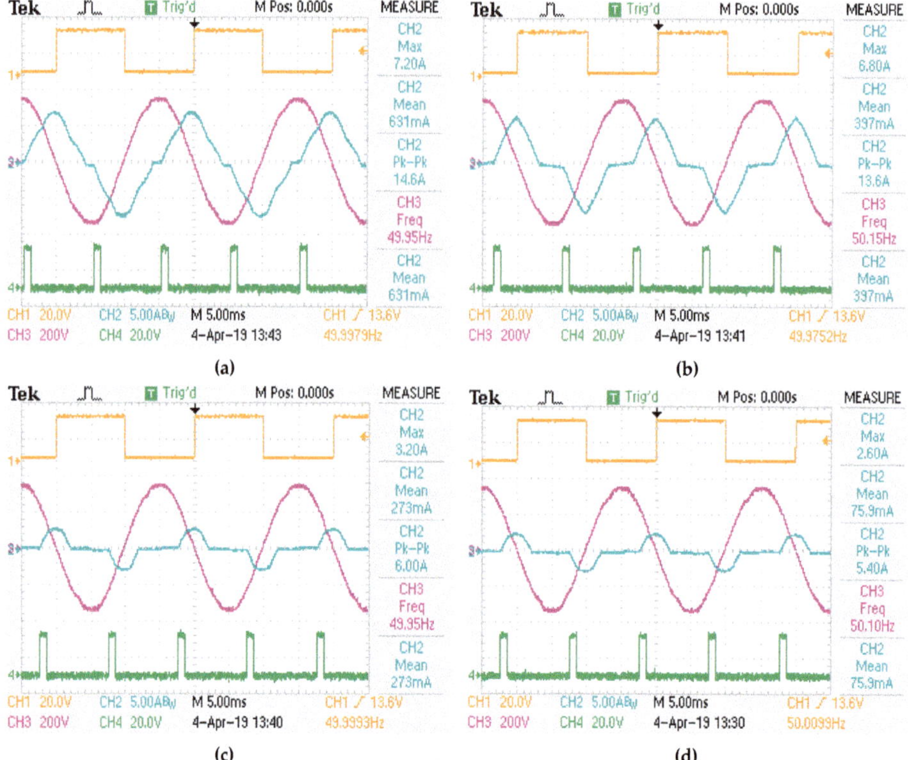

Figure 5. The waveforms of the utility-grid phase voltage (violet) and the reactor current (cyan) on thyristor firing angle (α): (**a**) 95° (**b**) 110° (**c**) 140° and (**d**) 160°. Voltage zero crossing is displayed in yellow, thyristor control signal in green.

Spectrum analysis was carried out employing an FFT toolbox by importing oscilloscope data into MATLAB/Simulink. The spectrums were obtained experimentally; therefore, harmonic 0 can appear because of measurement error or because the current curve is slightly asymmetric with respect to the time axis, i.e., some DC bias may exist. The asymmetry can be introduced by nonlinearity of our facility network, which can be caused by other devices powered from the same network. The spectrums of reactor current at various thyristor firing angles are presented in Figure 6, and the total harmonic distortion (THD) in Table 2. When a reactor consumes a large amount of reactive energy, which requires the current to flow through the reactor for practically the whole period, the current shape is distorted slightly (Figure 5a). However, for the reduction of the consumed reactive energy, it is necessary for the current to flow through the reactor only for part of the period, so the current shape distortion increases (Figure 5c,d). On the other hand, as the current through the reactor decreases, its effect on the overall distortion of the grid current also decreases.

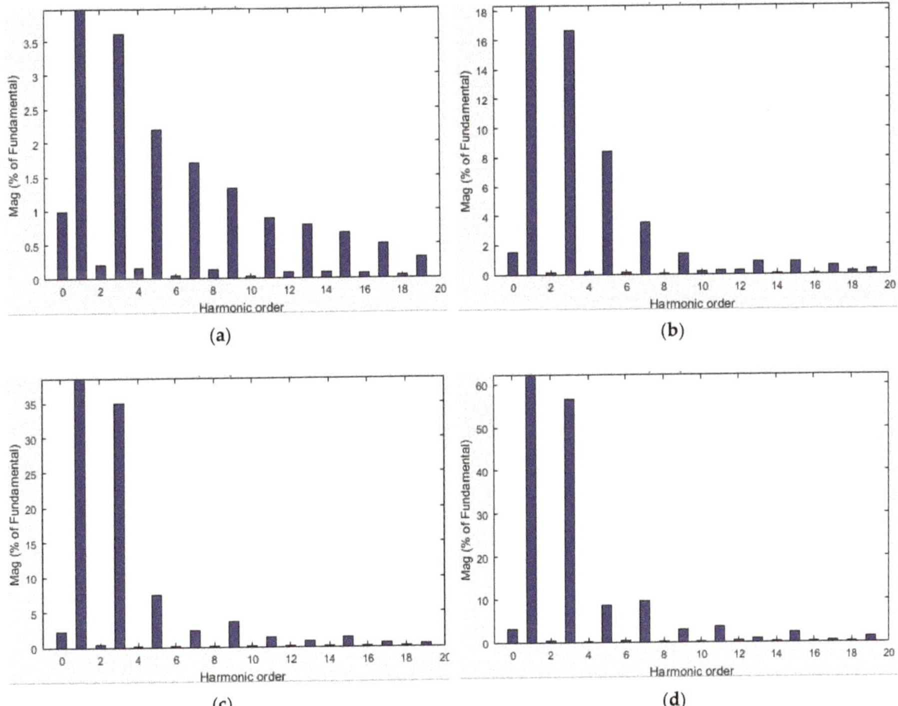

Figure 6. The spectrums of the reactor current at various thyristor firing angles: (**a**) 95° (**b**) 110° (**c**) 140° and (**d**) 160°. The frequency of the fundanental harmonic is 50 Hz.

Table 2. Total harmonic distortion of the single-cored three-phase air-gaped reactor current.

Thyristor Firing Angle (α)	Total Harmonic Distortion (THD), %
95°	5.2
110°	19.2
140°	36.1
160°	58.3

The next experiment was conducted to determine whether it is possible to control the consumed reactive power in every phase independently, using a single-cored three-phase air-gaped reactor. This is important because only the possibility of independent consumption of reactive power in each phase allows us to implement asymmetric compensation. The experiment was performed for the case when the firing angles of two phases were fixed: the firing angle of one phase was fixed at 165° (corresponds to minimal reactive power), while the firing angle of another phase at 99° (corresponds to maximal reactive power). The firing angle of the remaining phase was varied. The obtained dependences are presented in Figure 7. It is seen that variation of the firing angle of one phase does not just change the reactive power of the controlled phase but also influences the reactive power of phases with fixed firing angles.

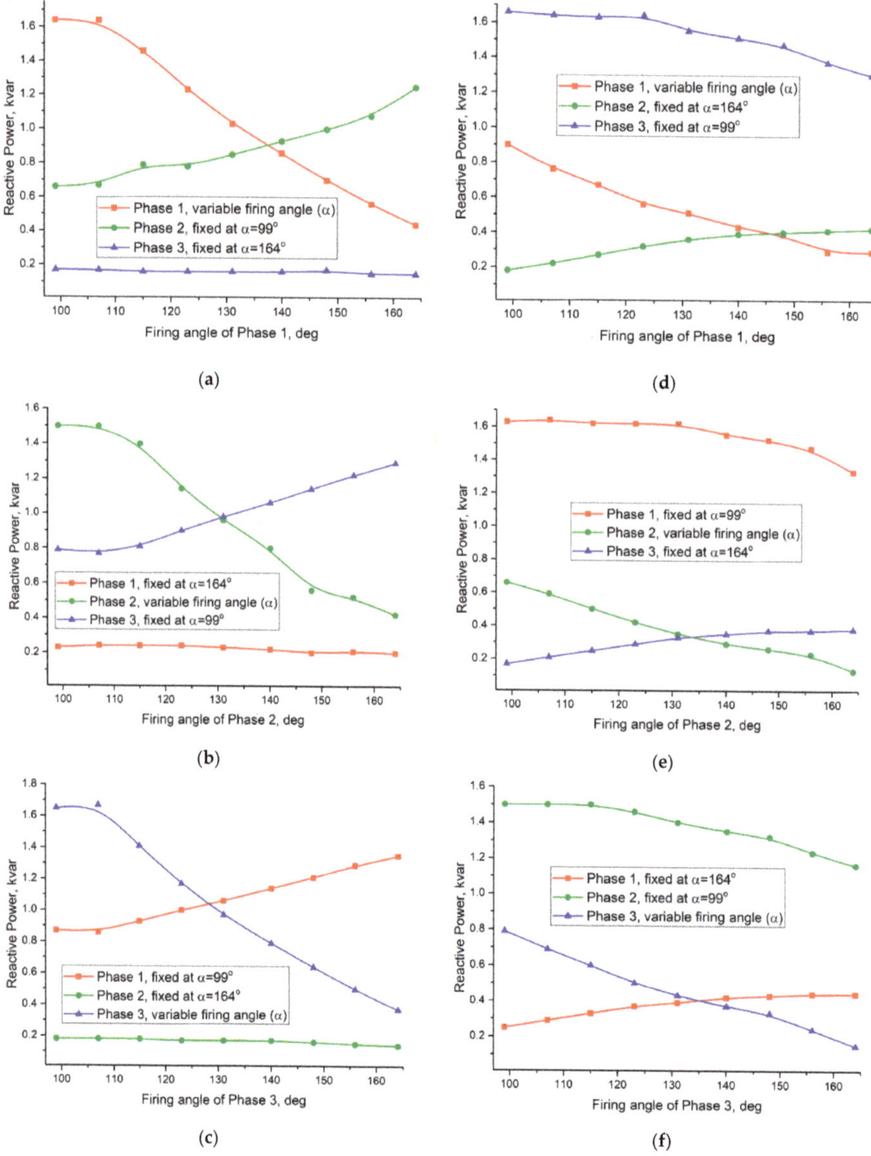

Figure 7. Dependences of reactive power consumed by the single-cored three-phase reactor on the firing angle when the firing angle of one phase is variable and the angles of the remaining two phases are fixed. The Firing angle is variable for Phase 1 (**a,d**), for Phase 2 (**b,e**), for Phase 3 (**c,f**).

The nature of reactive power dependences for the phases with the fixed firing angle depends on the phase sequence. For one sequence, the reactive power was influenced only in one of the phases with a fixed angle (Figure 7a–c). For another sequence, the reactive power of both phases with the fixed firing angles was affected (Figure 7d–f).

During the next experiment, the firing angles of two phases were varied and the angle of the remaining phase was fixed. The investigation was performed for the case when the fixed firing angle was set to $\alpha = 165$. Dependences of reactive power consumed by the single-cored three-phase reactor on firing angles of two phases are presented in Figure 8. It is seen that reactive power dependences of the phases with the variable firing angle strongly differ in spite of the fact that the firing angles of the thyristors are varied simultaneously. This happens due to one phase being influenced by another through the common core of the reactor.

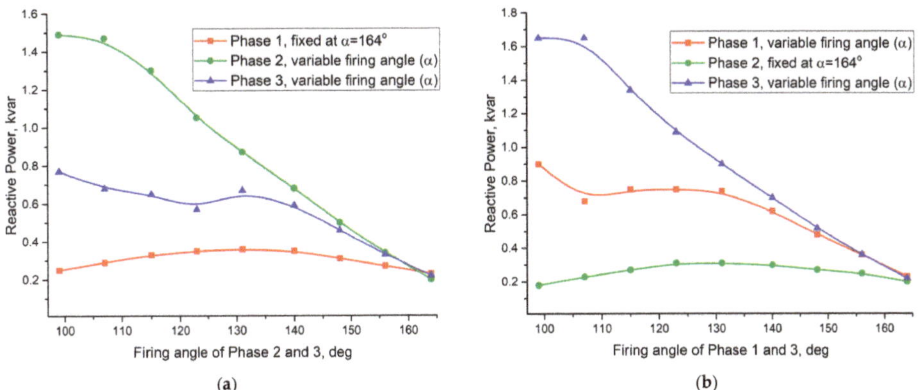

Figure 8. Dependences of reactive power consumed by the single-cored three-phase reactor on the firing angle when the firing angles of the two phases are variable and the angle of the remaining phase is fixed. The Firing angle is variable for Phase 1 (**a**), for Phase 2 (**b**).

Summarizing the obtained experimental investigation results, it can be concluded that it is impossible to control the reactive power in every phase independently using a compensator based on a single-cored three-phase reactor. This happens because phases influence each other through the common core of the reactor; therefore, separate reactors must be used for each phase for the asymmetric compensation of reactive power in a low-voltage utility grid.

3.2. Investigation of the Compensator Based on Separate Reactors for Every Phase

The structure and view of the designed air-gaped reactor for single phase are presented in Figure 9. Every single-phase reactor is capable of consuming 4.2 *kVAr* of reactive power, which corresponds to phase current RMS $I = 18.5$ A for the low-voltage utility-grid phase voltage $|U| = 230$ V. Total reactive power of all three reactors is 12.6 kVAr. The impedance of the reactor is $|Z| = \frac{|U|}{|I|} \approx 12.5\ \Omega$; the inductance $L = \frac{X_L}{\omega} \approx 40\ mH$. The required 40 mH inductance value was achieved by adjusting the reactor air gap. The parameters of the single-phase air-gaped reactor are presented in Table 3.

The TCR compensator based on three single-phase air-gaped reactors was investigated experimentally. Firstly, the reactive power dependences on the firing angle of thyristors, when firing angles in all three phases were changed simultaneously (in case of case of smooth symmetric compensation), were obtained (Figure 10). It is seen that the reactive power consumed by the single-phase reactors changes in all phases by the same law. The dispersion of the reactive power between individual phases was about 3%. The next experiment was performed in the same way as in the case of the TCR based on a single-cored three-phase air-gaped reactor, i.e., the firing angles of two phases were fixed: The firing angle of one phase was fixed at 165°, while the firing angle of another phase was fixed at 99°. The firing angle of the remaining phase was varied. The obtained dependences of reactive power consumption of every phase on the firing angle are presented in Figure 11. It is seen that the reactive power consumption of the phase with the variable firing angle has no impact on the reactive power consumption of the remaining two phases with the fixed firing angles. Therefore, this conclusion can be drawn: The employment of three single-phase reactors allows us to control the reactive power in every phase independently, and the compensator with three single-phase reactors is suitable for the smooth asymmetric compensation of reactive power in a low-voltage utility grid.

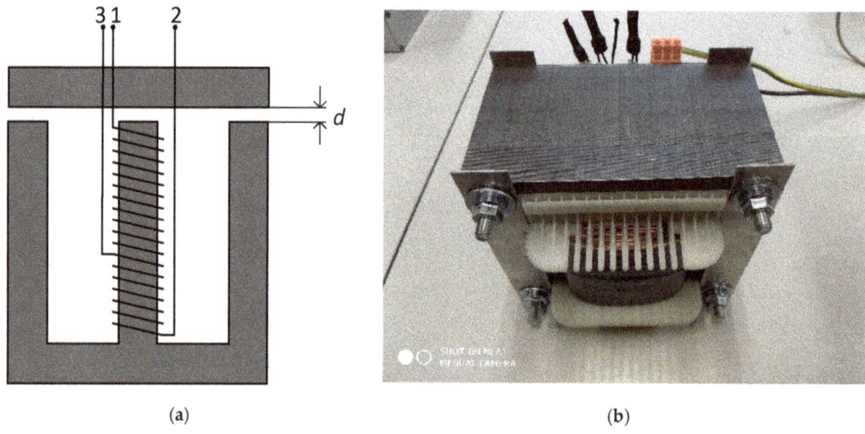

(a)   (b)

Figure 9. Design (**a**) and view (**b**) of the single-phase air-gaped reactor.

Table 3. Parameters of the single-phase air-gaped reactor.

Parameter	Value
Relative magnetic permeability of iron core (μ_I)	100
Number of turns of coil (N)	160
Winding area (S)	71.5 cm^2
Length of coil (l)	9.0 cm
Wire cross-section	3.1 mm^2
Inductance of coil at core air gap length $d = 0$	256 mH
Inductance of coil at $d = 5$ mm	40 mH
Inductance of coil at $d = 10$ mm	18 mH
Inductance of coil without core	2.6 mH

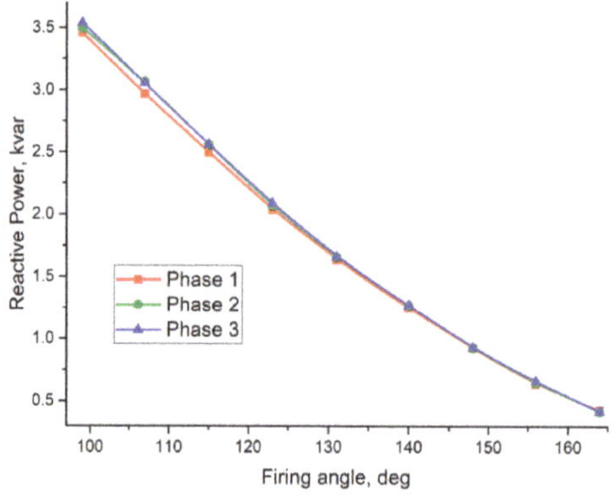

Figure 10. Dependencies of reactive power consumed by the single-phase air-gaped reactors on the firing angle of thyristors.

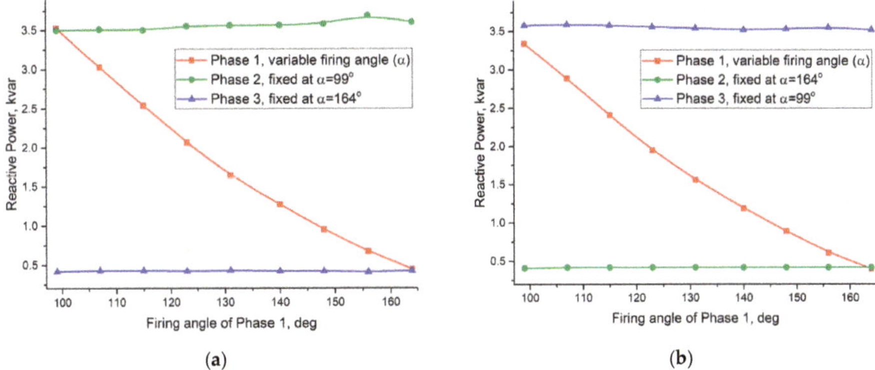

Figure 11. Dependences of reactive power consumed by each single-phase air-gaped reactor on the firing angle when the firing angles of the two phases are fixed and the angle of the remaining phase is variable. The Firing angle is variable for Phase 1 (**a**), for Phase 2 (**b**).

It should be mentioned that the investigation also covered the TCR compensator with Δ connection of single-phase reactor coils as well as Y-connection with an unconnected midpoint; however, the investigation results showed that these topologies of the TCR compensator are not suitable for asymmetric load compensation.

3.3. Efficiency of the TCR Compensator

The dependences of reactive and active power of single-cored three-phase and separate-phase air-gaped reactors were measured by applying a symmetric load (Figure 12). The efficiency of reactors was calculated as a ratio of reactive power to total power. Dependencies of reactor efficiency on the firing angle are given in Figure 13.

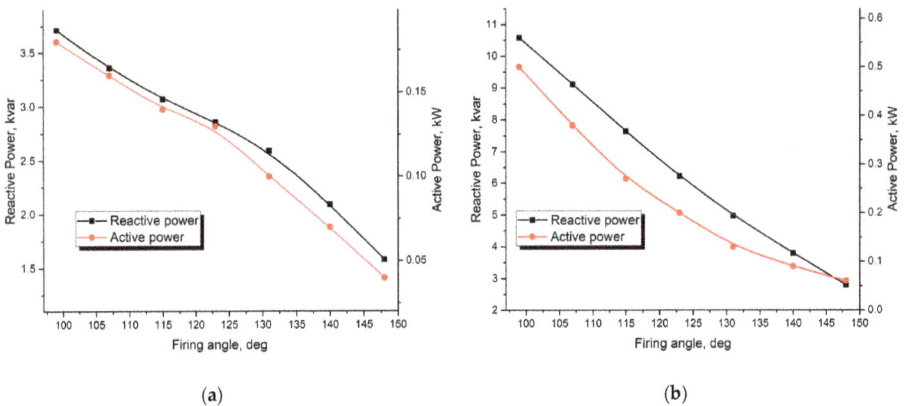

Figure 12. Dependences of reactive and active power consumed by (**a**) the single-cored three-phase reactor and (**b**) three single-phase air-gaped reactors.

It could be observed (Figure 13) that the efficiency of the reactors varies from 0.955 to 0.975 when power consumed by the reactors changes from the maximal to minimal value. It is seen that efficiency decreases with increasing of the reactive power (with decreasing of the thyristor firing angle). This appears due to the fact that as the reactive power increases, the reactor current increases, and as a consequence, the active reactor losses increase as well.

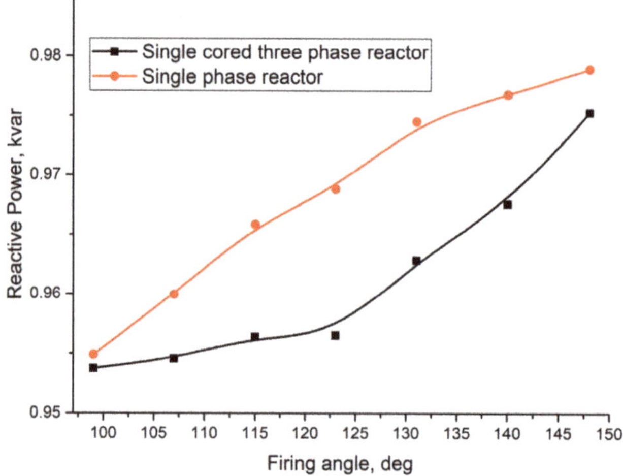

Figure 13. Dependencies of reactor efficiency on the firing angle.

4. Conclusions

1. TCR compensators, which typically are used in high- and medium-voltage utility grids, can be implemented in a low-voltage utility grid employing air-gapped reactors using a Y-connection connected to the neutral midpoint.

2. Variation of the thyristor firing angle of one phase of single-cored three-phase reactor does not just change the reactive power of the controlled phase but influences the reactive power of phases with fixed firing angles. This fact shows that it is impossible to control the reactive power in every phase independently using a TCR compensator based on a single-cored three-phase air-gaped reactor, i.e., a compensator with such a reactor is not suitable for the asymmetric compensation of reactive power.

3. Employment of three single-phase air-gaped reactors allows us to control the reactive power in every phase independently; therefore, a developed TCR compensator based on three single-phase reactors is suitable for smooth and asymmetric compensation of reactive power in a low-voltage utility grid.

4. Commutation of the reactor using thyristor switches does not introduce any high-frequency disturbances of the reactor current and grid voltage.

5. TCR compensator topologies with Δ connection of coils of single-phase reactors as well as Y-connection with unconnected midpoint are not suitable for asymmetric compensation of reactive power in a low-voltage utility grid.

6. The developed single-cored three-phase reactor and single-phase reactors are characterized by 0.955–0.975 efficiency.

Author Contributions: The results presented in this paper were obtained in the framework of a PhD thesis by M.Š., who was supervised by A.B. The experimental measurements performed at the Laboratory of Electronic Sytems, Center for Physical Sciences and Technology were supervised by M.Š., V.B., N.P. and A.D. The conceptualization was introduced by S.P., A.B., V.M. and E.B. All authors contributed to the writing of this paper as well as the analysis of the results. All authors have read and agreed to the published version of the manuscript.

Funding: The paper is funded by UAB Lietpak and UAB Navitus under Grant No. 3400-S510.

Acknowledgments: This work was supported, in part, by UAB Lietpak and UAB Navitus under Grant No. 3400-S510.

Conflicts of Interest: The authors declare no conflict of interest.

References

1. Yong, X.; Jiamei, D.; Shuangbao, M. Power flow control of a distributed generation unit in micro-grid. In Proceedings of the 2009 IEEE 6th International Power Electronics and Motion Control Conference, Wuhan, China, 17–20 May 2009; pp. 2122–2125.
2. Trentini, F.; Tasca, M.; Tomasin, S.; Erseghe, T. Reactive power compensation in smart micro grids: A prime-based testbed. In Proceedings of the 2012 IEEE International Energy Conference and Exhibition (ENERGYCON), Florence, Italy, 9–12 September 2012; pp. 909–914.
3. Deblecker, O.; Stevanoni, C.; Vallée, F. Cooperative control of multi-functional inverters for renewable energy integration and power quality compensation in micro-grids. In Proceedings of the 2016 International Symposium on Power Electronics, Electrical Drives, Automation and Motion (SPEEDAM), Anacapri, Italy, 22–24 June 2016; pp. 1051–1058.
4. Ilisiu, D.; Dinu, E.-D. Modern reactive power compensation for smart electrical grids. In Proceedings of the 2019 22nd International Conference on Control Systems and Computer Science (CSCS), Bucharest, Romania, 28–30 May 2019; pp. 353–357.
5. Bogónez-Franco, P.; Balcells, J.; Junyent, O.; Jordà, J. SVC model for voltage control of a microgrid. In Proceedings of the 2011 IEEE International Symposium on Industrial Electronics, Gdansk, Poland, 27–30 June 2011; pp. 1645–1649.
6. Balcells, J.; Bogónez-Franco, P. Voltage control in a LV microgrid by means of an SVC. In Proceedings of the IECON 2013-39th Annual Conference of the IEEE Industrial Electronics Society, Vienna, Austria, 10–13 November 2013; pp. 6027–6030.
7. Beck, Y.; Berkovich, Y.; Müller, Z.; Tlusty, J. A Matlab-Simulink model of network reactive power compensation based on binary switchable capacitors and thyristor-controlled reactor. In Proceedings of the 2016 IEEE International Conference on the Science of Electrical Engineering (ICSEE), Eilat, Israel, 16–18 November 2016; pp. 1–5.
8. Beck, Y.; Berlovich, Y.; Braunstein, A. A Matlab-Simulink model of AC grid with a FC-TCR and invariant control system for reactive power compensation. In Proceedings of the 2016 International Symposium on Power Electronics, Electrical Drives, Automation and Motion (SPEEDAM), Anacapri, Italy, 22–24 June 2016; pp. 1292–1297.
9. Dong, T.; Li, L.; Ma, Z. A combined system of APF and SVC for power quality improvement in microgrid. In Proceedings of the 2012 Power Engineering and Automation Conference, Wuhan, China, 18–20 September 2012; pp. 1–4.
10. Ali, Z.; Christofides, N.; Hadjidemetriou, L.; Kyriakides, E. Photovoltaic reactive power compensation scheme: An investigation for the Cyprus distribution grid. In Proceedings of the 2018 IEEE International Energy Conference (ENERGYCON), Limassol, Cyprus, 3–7 June 2018; pp. 1–6.
11. Arshad, A.; Lehtonen, M. Instantaneous active/reactive power control strategy for flicker mitigation under high PV penetration. In Proceedings of the 2018 IEEE PES Innovative Smart Grid Technologies Conference Europe (ISGT-Europe), Sarajevo, Bosnia and Herzegovina, 21–25 October 2018; pp. 1–6.
12. Stanelyte, D.; Radziukynas, V. Review of voltage and reactive power control algorithms in electrical distribution networks. *Energies* **2020**, *13*, 58. [CrossRef]
13. Gayatri, M.T.L.; Parimi, A.M.; Kumar, A.V.P. A review of reactive power compensation techniques in microgrids. *Renew. Sustain. Energy Rev.* **2018**, *81*, 1030–1036. [CrossRef]
14. Charalambous, A.; Hadjidemetriou, L.; Zacharia, L.; Bintoudi, A.D.; Tsolakis, A.C.; Tzovaras, D.; Kyriakides, E. Phase balancing and reactive power support services for microgrids. *Appl. Sci.* **2019**, *9*, 5067. [CrossRef]
15. Li, B.; Tian, X.; Zeng, H. A grid-connection control scheme of PV system with fluctuant reactive load. In Proceedings of the 2011 4th International Conference on Electric Utility Deregulation and Restructuring and Power Technologies (DRPT), Weihai, China, 6–9 July 2011; pp. 786–790.
16. Santacana, E.; Rackliffe, G.; Tang, L.; Feng, X. Getting Smart. *IEEE Power Energy Mag.* **2010**, *8*, 41–48. [CrossRef]
17. Bielskis, E.; Baskys, A.; Valiulis, G. Controller for the grid-connected microinverter output current tracking. *Symmetry* **2020**, *12*, 112. [CrossRef]
18. Bielskis, E.; Baskys, A.; Sapurov, M. Single stage microinverter based on two-switch DC-DC flyback converter. *Elektron. Ir Elektrotechnika* **2017**, *23*, 29–32. [CrossRef]

19. Pană, A.; Băloi, A.; Molnar-Matei, F. From the balancing reactive compensator to the balancing capacitive compensator. *Energies* **2018**, *11*, 1979. [CrossRef]

20. Pană, A.; Băloi, A.; Molnar-Matei, F. Iterative method for determining the values of the susceptances of a balancing capacitive compensator. *Energies* **2018**, *11*, 2742. [CrossRef]

21. Xiang-Qian, T.; Keqing, X.; Ming, S.; Xianhong, M. Reactive power and unbalance compensation with DSTATCOM. In Proceedings of the 2005 International Conference on Electrical Machines and Systems, Nanjing, China, 27–29 September 2005; Volume 2, pp. 1181–1184.

22. Fuchs, E.; Masoum, M.A.S. *Power Quality in Power Systems and Electrical Machines*; Academic Press: Cambridge, MA, USA, 2011; ISBN 978-0-08-055917-9.

23. Igbinovia, F.; Fandi, G.; Svec, J.; Muller, Z.; Tlusty, J. Comparative review of reactive power compensation technologies. In Proceedings of the 2015 16th International Scientific Conference on Electric Power Engineering, Ostrava, Czech Republic, 20–22 May 2015; pp. 2–7. [CrossRef]

24. Acha, E.; Agelidis, V.; Anaya-Lara, O.; Miller, T.J.E. *Power Electronic Control in Electrical Systems*; Newnes: Oxford, UK, 2002; ISBN 978-0-7506-5126-4.

25. Dixon, J.; Moran, L.; Rodriguez, J.; Domke, R. Reactive power compensation technologies: State-of-the-art review. *Proc. IEEE* **2005**, *93*, 2144–2164. [CrossRef]

26. Padiyar, R.K. *Facts Controllers in Power Transmission and Distribution*; New Age International (P) Ltd.: New Delhi, India, 2007; ISBN 978-81-224-2541-3.

27. Mathur, R.M.; Varma, R.K. *Thyristor-Based FACTS Controllers for Electrical Transmission Systems*; John Wiley & Sons: Hoboken, NJ, USA, 2002; ISBN 978-0-470-54668-0.

28. Tehrani, K.-A.; Capitaine, T.; Barrandon, L.; Hamzaoui, M.; Rafiei, S.M.R.; Lebrun, A. Current control design with a fractional-order PID for a three-level inverter. In Proceedings of the Proceedings of the 2011 14th European Conference on Power Electronics and Applications, Birmingham, UK, 30 August–1 September 2011; pp. 1–7.

29. Xu, Y.; Tolbert, L.M.; Kueck, J.D.; Rizy, D.T. Voltage and current unbalance compensation using a static var compensator. *IET Power Electron.* **2010**, *3*, 977–988. [CrossRef]

30. Chang, W.-N.; Liao, C.-H. Design and implementation of a STATCOM based on a multilevel FHB converter with delta-connected configuration for unbalanced load compensation. *Energies* **2017**, *10*, 921. [CrossRef]

31. PQC-Statcon. Available online: https://library.e.abb.com/public/2b588b8dd20ce996c1257a37003639e4/PQC-STATCON_Flyer.pdf (accessed on 6 November 2019).

32. Yan, J. *Handbook of Clean Energy Systems*, 1st ed.; John Wiley & Sons: Chichester, UK, 2015; Volume 6, ISBN 978-1-118-38858-7.

33. Panfilov, D.I.; ElGebaly, A.E. Modified thyristor controlled reactor for static VAR compensators. In Proceedings of the 2016 IEEE International Conference on Power and Energy (PECon), Melaka, Malaysia, 28–29 November 2016; pp. 712–717.

34. Khonde, R.S.; Palandurkar, M.V. Simulation model of thyristor controlled reactor. *Int. J. Eng. Res. Technol.* **2014**, *3*, 1692–1694.

35. Panfilov, D.I.; ElGebaly, A.E.; Astashev, M.G. Design and evaluation of control system for static VAR compensators with thyristors switched reactors. In Proceedings of the 2017 IEEE 58th International Scientific Conference on Power and Electrical Engineering of Riga Technical University (RTUCON), Riga, Latvia, 12–13 October 2017; pp. 1–6.

36. Mahapatra, S.; Goyal, A.; Kapil, N. Thyristor controlled reactor for power factor improvement. *Int. J. Eng. Res. Appl.* **2014**, *4*, 55–59.

37. Awad, F.; Mansour, A.; Elzahab, E. Thyristor controlled reactor with different topologies based on fuzzy logic controller. *Int. J. Eng. Res.* **2015**, *4*, 498–505. [CrossRef]

38. Panfilov, D.I.; ElGebaly, A.E.; Astashev, M.G. Topologies of thyristor controlled reactor with reduced current harmonic content for static VAR compensators. In Proceedings of the 2017 IEEE International Conference on Environment and Electrical Engineering and 2017 IEEE Industrial and Commercial Power Systems Europe (EEEIC / I CPS Europe), Milan, Italy, 6–9 June 2017; pp. 1–6.

39. Čerňan, M.; Tlustý, J. Study of the susceptance control of industrial static var compensator. In Proceedings of the 2015 16th International Scientific Conference on Electric Power Engineering (EPE), Kouty nad Desnou, Czech Republic, 20–22 May 2015; pp. 538–541.

40. Hong, H.; Wenmei, W.; Shaohua, X.; Min, T.; Chuanjia, H. Summary on reactive power compensation technology and application. In Proceedings of the 2nd International Conference on Intelligent Computing and Cognitive Informatics (ICICCI 2015), Singapore, 8–9 September 2015.

41. Farkoush, S.G.; Kim, C.-H.; Rhee, S.-B. THD reduction of distribution system based on ASRFC and HVC method for SVC under EV charger condition for power factor improvement. *Symmetry* **2016**, *8*, 156. [CrossRef]

42. Alkayyali, M.; Ghaeb, J. Hybrid PSO–ANN algorithm to control TCR for voltage balancing. *IET Gener. Transm. Distrib.* **2020**, *14*, 863–872. [CrossRef]

43. Panfilov, D.I.; Rozhkov, A.N.; Astashev, M.G.; Zhuravlev, I.I. Modern approaches to controlled static VAR compensators design. In Proceedings of the 2019 IEEE International Conference on Environment and Electrical Engineering and 2019 IEEE Industrial and Commercial Power Systems Europe (EEEIC/I CPS Europe), Genova, Italy, 11–14 June 2019; pp. 1–5.

44. Köse, A.; Irmak, E. Modeling and simulation of a static VAR compensator based on FC-TCR. In Proceedings of the 2016 IEEE International Conference on Renewable Energy Research and Applications (ICRERA), Birmingham, UK, 20–23 November 2016; pp. 924–927.

45. Rahmani, S.; Hamadi, A.; Al-Haddad, K.; Dessaint, L.A. A Combination of shunt hybrid power filter and thyristor-controlled reactor for power quality. *IEEE Trans. Ind. Electron.* **2014**, *61*, 2152–2164. [CrossRef]

46. Liberado, E.V.; Souza, W.A.; Pomilio, J.A.; Paredes, H.K.M.; Marafão, F.P. Design of static VAr compensator using a general reactive energy definition. In Proceedings of the International School on Nonsinusoidal Currents and Compensation 2013 (ISNCC 2013), Zielona Góra, Poland, 20–21 June 2013; pp. 1–6.

47. Tokiwa, A.; Yamada, H.; Tanaka, T.; Watanabe, M.; Shirai, M.; Teranishi, Y. New hybrid static VAR compensator with series active filter. *Energies* **2017**, *10*, 1617. [CrossRef]

48. Arab-Tehrani, K.; Colteu, A.; Rasoanarivo, I.; Michel-Sargos, F. Design a new high intensity magnetic separator with permanent magnets for industrial applications. *Int. J. Appl. Electromagn. Mech.* **2010**, *32*, 237–248. [CrossRef]

49. Topaloglu, I. Air gap optimization of iron core shunt reactors with discretely disturbed air gaps for UHV systems. In Proceedings of the International conference on engineering and natural science (ICENS 2016), Sarajevo, Bosnia and Herzegovina, 24–28 May 2016; pp. 1–6.

50. Wass, T.; Hörnfeldt, S.; Valdemarsson, S. The design and construction of a controllable reactor with a HTS control winding. *J. Phys. Conf. Ser.* **2006**, *43*, 873. [CrossRef]

51. Bielskis, E.; Baskys, A.; Sapurov, M. Impact of transformer design on flyback converter voltage spikes. *Elektron. Ir Elektrotechnika* **2016**, *22*, 58–61. [CrossRef]

52. Šapurov, M.; Bielskis, E.; Bleizgys, V.; Dervinis, A. Stepless compensator of reactive power. *Moksl. Liet. Ateitis Sci. Future Lith.* **2020**, *12*. [CrossRef]

Article

Direct Power Compensation in AC Distribution Networks with SCES Systems via PI-PBC Approach

Walter Gil-González [1,*], **Federico Martin Serra** [2], **Oscar Danilo Montoya** [1,3],
Carlos Alberto Ramírez [4] and **César Orozco-Henao** [5]

[1] Laboratorio Inteligente de Energía, Universidad Tecnológica de Bolívar, km 1 vía Turbaco,
    Cartagena 131001, Colombia; omontoya@utb.edu.co
[2] Laboratorio de Control Automático (LCA), Universidad Nacional de San Luis,
    Villa Mercedes 5730, Argentina; fmserra@unsl.edu.ar
[3] Facultad de Ingeniería, Universidad Distrital Francisco José de Caldas, Carrera 7 No. 40B - 53,
    Bogotá D.C 11021, Colombia
[4] Facultad de Ciencias Básicas, Universidad Tecnológica de Pereira. AA: 97, Pereira 660003, Colombia;
    caramirez@utp.edu.co
[5] Electrical and Electronic Engineering Department, Universidad del Norte, Barranquilla 080001, Colombia;
    chenaoa@uninorte.edu.co
*   Correspondence: wjgil@utp.edu.co

Received: 18 February 2020; Accepted: 3 April 2020; Published: 23 April 2020

Abstract: Here, we explore the possibility of employing proportional-integral passivity-based control (PI-PBC) to support active and reactive power in alternating current (AC) distribution networks by using a supercapacitor energy storage system. A direct power control approach is proposed by taking advantage of the Park's reference frame transform direct and quadrature currents (i_d and i_q) into active and reactive powers (p and q). Based on the open-loop Hamiltonian model of the system, we propose a closed-loop PI-PBC controller that takes advantage of Lyapunov's stability to design a global tracking controller. Numerical simulations in MATLAB/Simulink demonstrate the efficiency and robustness of the proposed controller, especially for parametric uncertainties.

Keywords: distribution networks; direct power control; global tracking controller; passivity-based control; supercapacitor energy storage system

---

1. Introduction

Supercapacitors are promising energy storage technologies with high energy density and high charging/discharging capabilities [1]. They allow for the storage of electrical energy in electric fields, increasing their efficiency in comparison with mechanical or chemical devices [2]. Supercapacitor energy storage (SCES) and superconducting magnetic energy storage (SMES) are the only two devices that store energy in the form of electromagnetic fields [3]. Nevertheless, SCES systems are preferred because they work with voltage source converters [1], which are common and represent the cheapest option when compared with the current source converters in SMES applications [4]. Additionally, another advantage of using SCES over SMES systems is that the former does not require special thermal covers, such as those needed for cooling systems based on liquid hydrogen, helium, or nitrogen [3]. This increases the acquisition, installation, and maintenance of SMES systems.

The integration of SCES systems via voltage source converters allows controlling the active and reactive power independently by formulating a dynamic model in any reference frame (i.e., time domain or *abc*, Clark's or $\alpha\beta$, and Park's or *dq* reference frames, respectively) [5]. In addition, the dynamical model of the SCES system exhibits a nonlinear structure that makes it necessary to propose nonlinear controllers to deal with its operational goals. Multiple controllers for SCES systems integration in electrical networks have been proposed such as feedback linearization methods [6], interconnection and damping PBC approaches [3,5], linear matrix inequalities [7], classical proportional-integral controls [8], adaptive predictive control [9], or proportional-integral passivity-based control (PI-PBC) methods [2], that typically control active and reactive power in an indirect form by controlling the currents on the AC side of the converter. Two interesting approaches based on IDA-PBC and PI-PBC approaches have been reported by [10,11] to control SMES and SCES systems in autonomous applications of single-phase microgrids; these PBC approaches take the advantages of the port-Hamiltonian modeling to propose asymptotically stable controllers of Lyapunov. Authors confirm that single-phase converters allow the control of active and reactive power independently; nevertheless, the main disadvantage is the dependence on the parameters of the control laws, which complicates their application over systems with parametric uncertainties.

Note that PBC approaches are preferred to operate SCES systems because their dynamic models exhibit a port-Hamiltonian (pH) structure in open-loop, which is a suitable structure that is employed in PBC designs. Because it allows proposing closed-loop control structures that can guarantee stable operation in the sense of Lyapunov [12]. Even though the PI-PBC method has been previously presented for SCES systems by [2], who presented a direct power control structure, we proposed a robust parametric approach that avoids prior knowledge on the system parameters (inductance, resistance, and supercapacitor values) and does not require the solution of additional differential equations. This is a gap that is yet to be solved in specialized literature for SCES applications in AC distribution networks. In addition, the main advantage of using direct power control is that the state variables to design the controller are directly active and reactive power. Making more suitable the assignation of the references to control these variables in generation or load compensation applications, while classical approaches work with currents as state variables requiring additional steps regarding active and reactive power control.

The remainder of this paper is organized as follows: Section 2 presents a complete dynamical formulation of the SCES system interconnected to AC grids with a series resistive-inductive filter. In addition, we present its pH intrinsic formulation and its transformation from the current structure to the power ones. Section 3 presents the structure of the proposed PI-PBC approach, highlighting its independence regarding the filter parameters and provides a general proof to guarantee asymptotic convergence. Section 4 reports the general control structure as a function of active and reactive power measures as well as the physical constraints related to the integration of SCES in distribution networks. Section 5 presents the test system, simulating conditions, and numerical results with their corresponding analysis and discussion. Section 6 details the main conclusions derived from this research.

2. Dynamical Modeling

To obtain the dynamical representation of the SCES system integrated with VSC, only Kirchhoff's laws and the first Tellegen's theorem is required. Here, we suppose that all the variables were transformed from the *abc* reference frame into Park's reference frame [5].

$$l\frac{d}{dt}i_d = -ri_d + \omega l i_q + v_{sc}u_d - v_d, \tag{1}$$

$$l\frac{d}{dt}i_q = -ri_q - \omega l i_d + v_{sc}u_q - v_q, \tag{2}$$

$$c_{sc}\frac{d}{dt}v_{sc} = -i_d u_d - i_q u_q, \tag{3}$$

where l and r are the series inductance and resistance of the AC filter (transformer), respectively; c_{sc} is the capacitance value of the SCES system; i_d and i_q represent the direct- and quadrature-axis currents, respectively, while v_d and v_q are their corresponding voltages; v_{sc} is the voltage in the terminals of the supercapacitor and, u_d and u_q are the modulation indexes of the converter that work as control inputs. Note that dynamical Model (1)–(3) is a sub-actuate control system because there are $m = 2$ control variables and $n = 3$ states.

Remark 1. *The dynamical Model (1)–(3) exhibits a non-affine port-Hamiltonian structure as follows:*

$$\mathcal{D}\dot{x} = [\mathcal{J}(u) - \mathcal{R}]x + \zeta, \tag{4}$$

where $\mathcal{D} \in \mathbb{R}^{n \times n}$ is the inertia matrix, which is diagonal and positive definite, $\mathcal{J} \in \mathbb{R}^{n \times n}$ and $\mathcal{R} \in \mathbb{R}^{n \times n}$ are the interconnection and damping matrices, such that $\mathcal{J}$ is skew-symmetric and $\mathcal{R}$ is diagonal and positive semidefinite; $x \in \mathcal{R}^n$ and $u \in \mathcal{R}^m$ are the state and control vectors; and $\zeta \in \mathcal{R}^n$ corresponds to the external input vector.

Lemma 1. *The dynamical Model (1)–(3) can be transformed into a direct-power control (DPC) model preserving its non-affine port-Hamiltonian structure by defining the following variables as recommended in [2]: $p = v_d i_d$, $q = -v_d i_q$, and $v = v_d v_{sc}$, where $v_q = 0$ because the PLL is referred to the direct-axis.*

Proof. To obtain a DPC model, let us suppose that the v_d and v_q signals are obtained by implementing a phase-looked loop (PLL), such that they (i.e., v_d and v_q) are constants and well known. Now, if we derive the active and reactive power components, then,

$$\frac{d}{dt}p = v_d \frac{d}{dt}i_d, \tag{5a}$$

$$\frac{d}{dt}q = -v_d \frac{d}{dt}i_q, \tag{5b}$$

where, if we substitute Equation (1) and (2) considering that $v = v_d v_{sc}$, the following result is obtained

$$l\frac{d}{dt}p = -rp - \omega lq + vu_d - v_d^2, \tag{6a}$$

$$l\frac{d}{dt}q = -rq + \omega lp - vu_q. \tag{6b}$$

Note that if we multiply Equation (3) by v_d and rearrange some terms, then

$$c_{sc}\frac{d}{dt}v = -pu_d + qu_q. \tag{7}$$

Finally, when expressions Equations (5) to (7) are rearranged in the form of Equation (4), the proof is completed with

$$\mathcal{D} = \begin{pmatrix} l & 0 & 0 \\ 0 & l & 0 \\ 0 & 0 & c_{sc} \end{pmatrix}, \mathcal{J}(u) - \mathcal{R} = \begin{pmatrix} -r & -\omega l & u_d \\ \omega l & -r & -u_q \\ -u_d & u_q & 0 \end{pmatrix}$$

$$x = \begin{bmatrix} p & q & v \end{bmatrix}^T, \quad \zeta = \begin{bmatrix} -v_d^2 & 0 & 0 \end{bmatrix}^T.$$

$\square$

Remark 2. *Considering the advantages of the pH formulation exhibited by the DPC controller, an appropriate controller to alleviate the active and reactive power oscillations with the SCES systems is the passivity-based control approach because it takes advantage of the pH model to design an asymptotically stable controller in the sense of Lyapunov.*

3. Passivity-Based Control Design

PBC control is a powerful nonlinear control technique that allows designing stable controllers by taking advantage of pH modeling. The main objective of the PBC is to find a set of control laws that help to preserve the pH structure of the dynamical model in closed-loop to guarantee stability in the sense of Lyapunov, using energetic modeling [12]. There are different approaches based on the PBC theory; the most known approach corresponds to the interconnection and damping assignment (IDA-PBC) because it allows working with linear, nonlinear, and non-affine dynamical systems [13]. Nevertheless, there is also an interesting alternative that includes the well-known advantages of the PI actions in the PBC design, which produces a stable PI-PBC approach that guarantees stability in closed-loop operation; however, this approach is only applicable in nonlinear systems with a bilinear structure such as the case of the dynamical models of the power electronic converters [2]. In the next subsection, we will present the general PI-PBC design for bilinear systems.

3.1. Bilinear Representation

To develop a controller based on the PI-PBC approach, let us make the following definition.

Definition 1. *An admissible equilibrium point $(x^\star)$ exists for non-linear dynamical Model (4) if its variables are represented in Park's reference frame (direct- and quadrature-axis), such that*

$$0 = [\mathcal{J}(u^\star) - \mathcal{R}]x^\star + \zeta, \tag{8}$$

for some constant control input $u^\star$.

Considering the definition of the equilibrium point, now we use some auxiliary variables to develop a PI-PBC controller as follows: $\tilde{u} = u - u^\star$ and $\tilde{x} = x - x^\star$, where $\tilde{x}$ and $\tilde{u}$ represent the error of the state variables and control inputs.

Note that if we subtract Equation (8) from (4), the following result is obtained

$$\mathcal{D}\dot{\tilde{x}} = [\mathcal{J}(u)x - \mathcal{J}(u^\star)x^\star] - \mathcal{R}\tilde{x}. \tag{9}$$

To simplify Equation (9), let us define the property related to bilinear systems as follows:

Definition 2. *The matrix product $\mathcal{J}(u)x$ has a bilinear structure if it can be separated as a sum as follows*

$$\mathcal{J}(u)x = \mathcal{J}_0 + \sum_{i=1}^{m} \mathcal{J}_i x u_i, \tag{10}$$

where $\mathcal{J}_0$ and $\mathcal{J}_i$ are constant matrices with skew-symmetric structure.

Now, if we consider the Definition 2 in Equation (9) and make some algebraic manipulations, then the below result is obtained:

$$\mathcal{D}\dot{\tilde{x}} = [\mathcal{J}_0 - \mathcal{R}]\tilde{x} + \sum_{i=1}^{m} \mathcal{J}_i \tilde{x} u_i + \sum_{i=1}^{m} \mathcal{J}_i x^\star \tilde{u}_i. \tag{11}$$

Remark 3. *Expression (11) is the essential structure to design PI-PBC controllers for bilinear systems, as demonstrated in [14].*

3.2. Lyapunov's Requirements for Stability Analysis

To guarantee that the dynamical System (11) is stable in the sense of Lyapunov for the equilibrium point $\tilde{x} = 0$, i.e., $x = x^\star$, let us define a candidate Lyapunov function $\mathcal{V}(\tilde{x})$ with hyperboloid structure as presented below

$$\mathcal{V}(\tilde{x}) = \frac{1}{2}\tilde{x}^T \mathcal{D}\tilde{x}. \tag{12}$$

Observe that $\mathcal{V}(\tilde{x})$ meets the first two conditions of the Lyaponov's stability theorem, i.e., $\mathcal{V}(0) = 0$, and $\mathcal{V}(\tilde{x}) > 0$, $\forall \tilde{x} \neq 0$. In addition, if we take the temporal derivative of Equation (12) and substitute Equation (11), the following result is obtained:

$$\dot{\mathcal{V}}(\tilde{x}) = \tilde{x}^T \mathcal{D}\dot{\tilde{x}} = -\tilde{x}^T \mathcal{R}\tilde{x} + \sum_{i=1}^{m} \tilde{x}^T \mathcal{J}_i x^\star \tilde{u}_i, \tag{13}$$

which implies guarantee in stability, if the second term on the right hand side of Equation (13) is negative definite or at least negative semidefinite. To simplify this expression, let us use the input–output relation $\tilde{u} \to \tilde{y}$ being $\tilde{y}$ the passive output as follows

$$\dot{\mathcal{V}}(\tilde{x}) \leq \tilde{y}^T \tilde{u}, \tag{14}$$

where $\tilde{y}_i = \tilde{x}^T \mathcal{J}_i x^\star$.

Note that Expression (14) can help us to design a stable controller if and only if the set of control inputs $\tilde{u}$ is selected such that this expression is always negative semidefinite. These characteristics are presented in the next section using a PI controller.

3.3. PI-PBC Design

To obtain a general control law to guarantee closed-loop stability in the sense of Lyapunov, let us employ the following PI control structure

$$\tilde{u} = -\mathcal{K}_p \tilde{y} + \mathcal{K}_i z, \tag{15a}$$

$$\dot{z} = -\tilde{y}, \tag{15b}$$

where $\mathcal{K}_p \succ 0$ and $\mathcal{K}_i \succ 0$ are the proportional and integral gain matrices and z is an auxiliary vector of variables related to the integral action.

To prove stability with the control law defined by Equation (15), let us modify the candidate Lyapunov function in Equation (12) as follows

$$\mathcal{W}(\tilde{x}, z) = \mathcal{V}(\tilde{x}) + \frac{1}{2}(z - z_0)^T \mathcal{K}_i (z - z_0), \tag{16}$$

with $z_0 = \mathcal{K}_i^{-1} u^\star$ and its derivative is

$$\dot{\mathcal{W}}(\tilde{x}, \tilde{y}) = -\tilde{x}^T \mathcal{R}\tilde{x} - \tilde{y}^T \mathcal{K}_p \tilde{y} \leq 0, \tag{17}$$

which proves that the control input in Equation (15) guarantees stability in the sense of Lyapunov for closed-loop operation. Observe that in Equation (16), we consider that $\mathcal{K}_i = \mathcal{K}_i^T$.

Remark 4. *The control input Equation (15) can guarantee asymptotic stability in the sense of Lyapunov as proved in [15] by referring to Barbalat's lemma [14].*

4. Control Structure and Physical Constraint

This section presents the mathematical structure of the control laws for active and reactive power support with SCES systems, as well as the physical constraint that imposes the interconnection of a supercapacitor for energy storage applications.

4.1. Control Law

The presented PI-PBC approach can deal with parametric uncertainties in the SCES system when it is modeled using a direct power formulation (see Model (7) and (8)). For its analysis, let us present the general control inputs obtained from Equation (15) as follows

$$u_d = u_d^\star + k_{p1}\left(v^\star\left(p - p^\star\right) - p^\star\left(v - v^\star\right)\right) + k_{i1}\int\left(v^\star\left(p - p^\star\right) - p^\star\left(v - v^\star\right)\right)dt, \tag{18a}$$

$$u_q = u_q^\star + k_{p2}\left(q^\star\left(v - v^\star\right) - v^\star\left(q - q^\star\right)\right) + k_{i2}\int\left(q^\star\left(v - v^\star\right) - v^\star\left(q - q^\star\right)\right)dt, \tag{18b}$$

where k_{p1} and k_{p2} are the proportional gains, and k_{i1} and k_{i2} are the integral gains, respectively.

Remark 5. *The components $u_d^\star$ and $u_q^\star$ in Equation (18) can be neglected as recommended in [16] because they can be considered as constant values to calculate integral actions.*

It is important to mention that the control inputs of Equation (18) can remain robust to parametric uncertainties because they do not depend on any parameter of the system. This implies that small variations in these values (e.g., l, r, and c_{sc}) will not compromise the dynamical performance of the SCES system.

Note that in Equation (18), the value of $v^\star = v_d v_{sc}^\star$ needs to control the active and reactive power interchange between the SCES system and the grid (the values of $p^\star$ and $q^\star$ are defined by the designer because the main interest in SCES applications corresponds to control active and reactive power independently). Therefore, it is necessary to know $v_{sc}^\star$ to apply the controller. We start from the energy function of the SCES to compute $v_{sc}^\star$, as follows:

$$W_{sc}^\star = \frac{1}{2}C_{sc}v_{sc}^{\star\,2} \rightarrow \dot{W}_{sc}^\star = p_{sc}^\star = C_{sc}\dot{v}_{sc}^{\star\,2}, \tag{19}$$

and the relation between the active power of SCES and VSC can be approximated to $p_{sc}^\star = -p^\star$. Hence, $v_{sc}^\star$ can be given by

$$v_{sc}^{\star\,2} = \frac{1}{C_{sc}}\int -p^\star dt \rightarrow v_{sc}^\star = K_i\sqrt{\int_0^t -p^\star dt}, \tag{20}$$

with $K_i > 0$.

It is important to mention that the stability proof shown in Section 3.3 may be compromised by replacing Equation (20) into (18), which is only valid when $v_{sc}^\star$ is constant. Therefore, we adopted the time-scale separation assumption between the outer-loop ($v_{sc}^\star$) and the inner PI-PBC described in [17,18]. Interestingly, the assumption deals with the possible lost stability by only adjusting the integral gain in Equation (20).

4.2. Physical Operative Constraint

The energy storage capability of a SCES system is limited by the energy capabilities of the supercapacitor as well as for the admissible voltages in its terminals.

Figure 1 presents the dynamic behavior of the total energy stored in the SCES. In this figure, there are three critical points called O, P, and Q. Point O presents the minimum voltage value permissible in the supercapacitor terminals ($v_{sc}^{\min}$) that produce the value of the admissible minimum energy stored

(E_{min}); this voltage value is the lower bound in the SCES operation to guarantee the controllability of the closed-loop system. Additionally, point Q is the upper bound of the energy storage variable, which reaches the maximum permissible energy stored in the supercapacitor device ($v_{sc}^{max} \leftrightarrow E_{max}$), whereas point P represents some operating points between extreme points of O and Q.

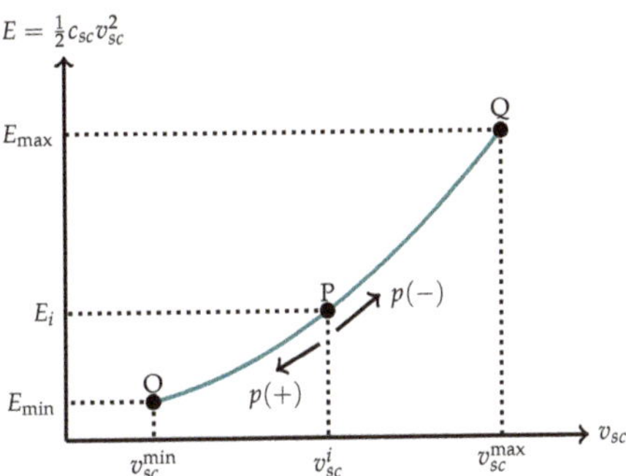

Figure 1. Behavior of the energy stored in the Supercapacitor energy storage (SCES).

Observe that point P allows positive or negative active power references. In the case in which p is positive, the energy stored decreases from point P to point O. In the case in which p is negative, the energy stored increases from point P to point Q. When point P is at the extreme points, we can conclude that if it is at point O, the reference for p can only be negative or zero, and if it is at point Q, it must be positive or zero. These are necessary conditions to preserve the useful life in the SCES.

5. Test System and Simulation Scenarios

5.1. The System Under Study

A low-voltage microgrid is used to test the DPC model of the SCES system. The test system is depicted in Figure 2, which contains two SCES systems. Additionally, it has a wind power generator and unbalanced loads to generate power oscillations in the system, thus allowing SCES to compensate for oscillations. The test system data can be consulted in [2].

To assess the capability and robustness of the proposed controller, the controller is compared with interconnection and damping assignment IDA-PBC presented in [19]. The IDA-PBC was performed under the $\alpha\beta$ reference frame. This implies that a PLL does not need to be implemented. However, it requires the derivatives of the desired values and the parameters of the system to be employed. This entails that the implementation of the IDA-PBC is more complicated than the proposed controller.

5.2. Simulation Scenarios

The SCES system can be used in various forms as support of active and reactive power oscillations in distributed generation applications with the wind turbine generator or can compensate for power oscillations provided by unbalanced loads [19]. According to this, in this study, two scenarios are designed to assess the performance and robustness of the SCES system using the DPC model and the proposed controller. The scenarios are as follows:

- **Scenario 1 (S1):** Check the proposed controller to manage the active and reactive power independently in the SCES system.
- **Scenario 2 (S2):** Evaluate the performance of the proposed controller applied to the SUCCESS system using the DPC model to relieve the oscillations of active and reactive power in the microgrid. This scenario employs a wind power generator located at bus 2, which provides active power and absorbs the reactive power shown in Figure 3. Additionally, the test system has two

demands (see DL1 and DL2 loads in Figure 2), which draw active and reactive power, as depicted in Figure 4.

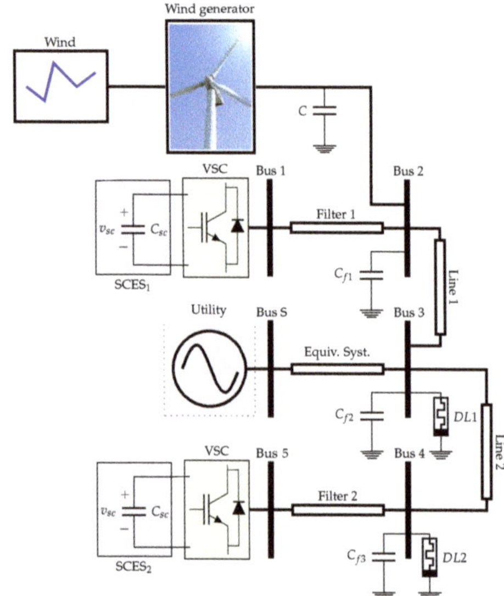

Figure 2. Test system configuration.

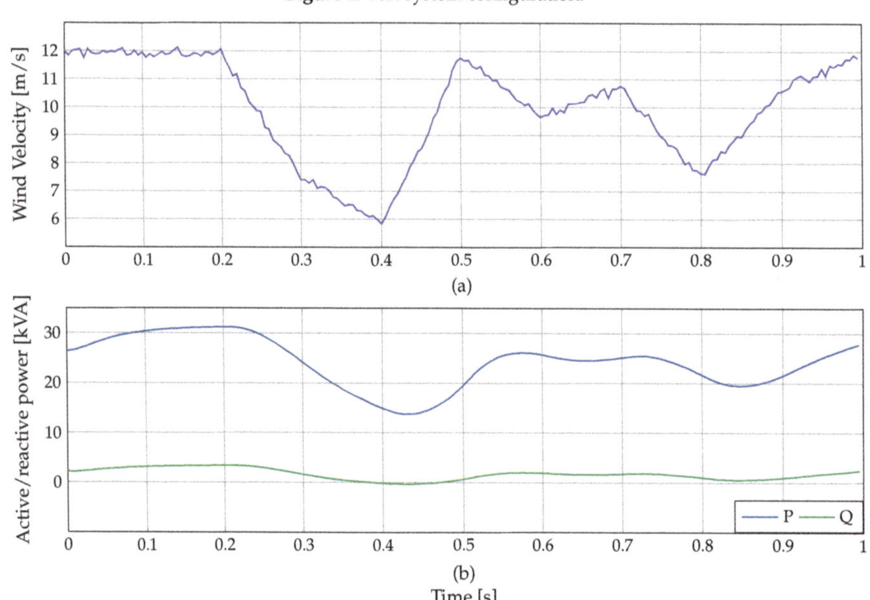

Figure 3. Wind generator: (**a**) wind profile and (**b**) active power provided and reactive power absorbed.

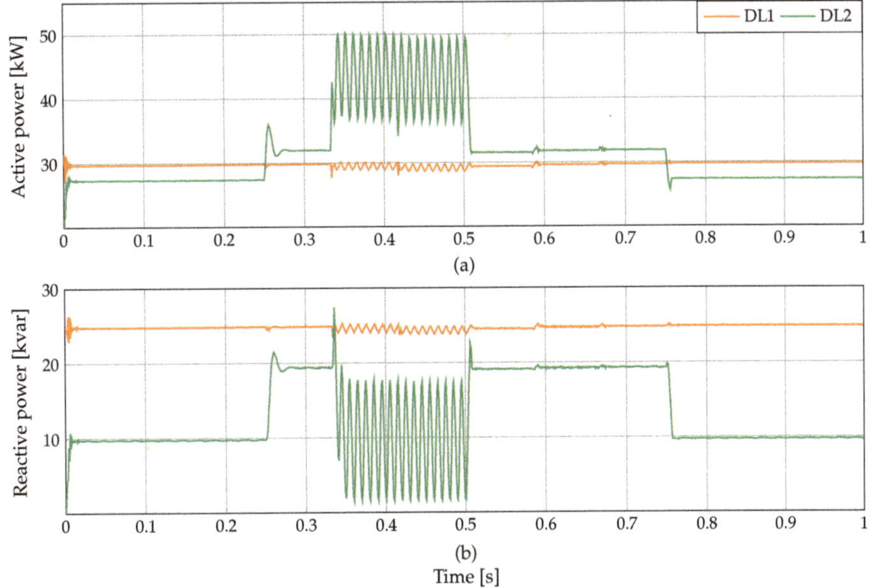

Figure 4. DL1 and DL2 loads: (**a**) active power demanded and (**b**) reactive power demanded.

It is important to mention that these scenarios have been designed arbitrarily to validate the accuracy and robustness of the DPC model and the proposed controller.

6. Results

The test system (see Figure 2) was implemented in MATLAB/Simulink and was executed on a desk-computer INTEL(R) Core(TM) i7–7700 CPU, 3.60 GHz, 8 GB RAM with 64-bits Windows 10 Professional by using MATLAB2019b. The switching frequency for the VSCs was fixed at 5 kHz. The parameters for the PI-PBC approach were tuned using the diagonal method developed in [20], which is suitable when the connection between the internal control of the model and the PI controller is lost, as presented with the proposed controller in this paper. The PI controller is computed, as follows:

$$k_{p_{12}} = \frac{\ln(9)L}{t_s}, k_{i_{12}} = \frac{\ln(9)R}{t_s},$$ (21)

where L and R are values of the inductance and resistance of the AC filter and $t_s = 10^{-4}$ is the desired settling time.

6.1. Scenario 1

This scenario analyzes the performance of the PI-PBC to control the active and reactive power regardless of the SCES system applied to the DPC model. Here, we select arbitrary values for active and reactive power references to demonstrate the ability of the proposed controller. Figure 5 shows the dynamic behavior at the DC side of the SCES system and their active and reactive power outputs.

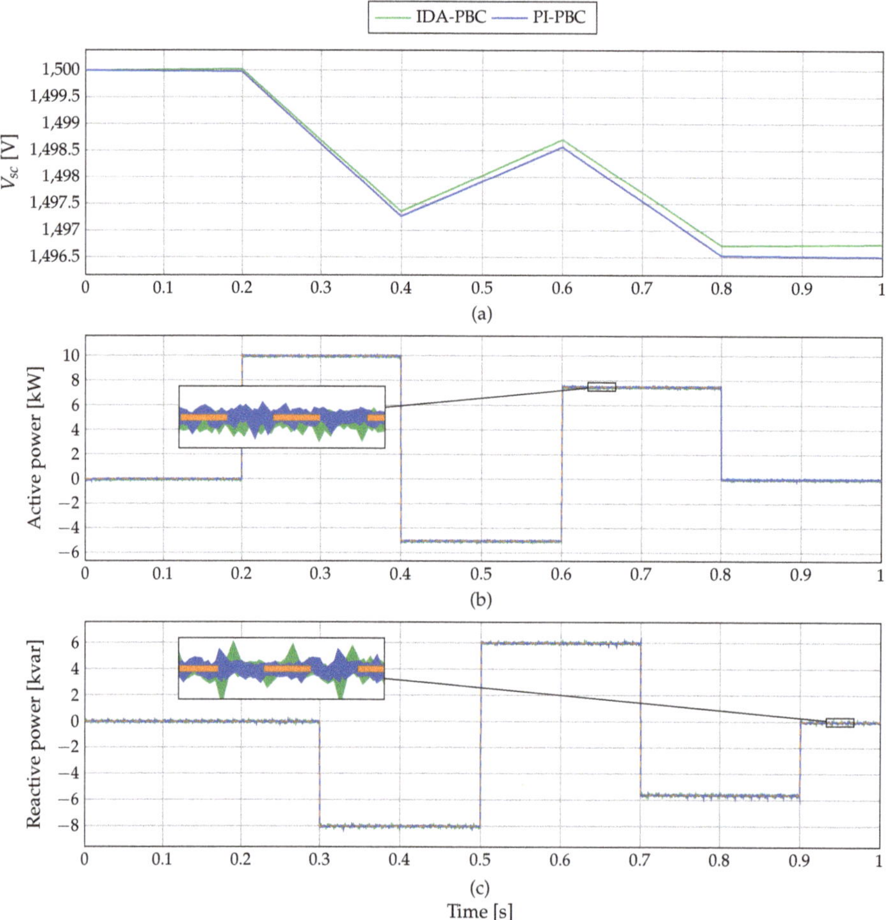

Figure 5. Dynamic behavior of the SCES system for S1: (**a**) supercapacitor voltage; (**b**) active power provided; and (**c**) reactive power absorbed.

Figure 5 shows that both controllers present a similar dynamic behavior. Nevertheless, PI–PBC presents a better performance compared with the IDA-PBC because it shows lower active and reactive power ripples. The ripples for the proposed controller are approximately 80 W and 85 var for the active and reactive power, respectively. In contrast, the ripples are about 100 W and 115 var when the IDA-PBC is implemented, respectively.

6.2. Scenario 2

This scenario investigates the ability of the SCES system using the DPC model to compensate for the power oscillations in the test system. Therefore, two SCES systems are considered, located at buses 1 and 5 (see Figure 2). The first SCES system relieves the power oscillation introduced by the wind power generator at bus 2. Here, we assume that the SCES system must keep active power at 28 kW and supply all the requirements of the reactive power of the generator. In contrast, the second SCES system compensates for the power oscillations provided by DL2 demands at bus 4, maintaining its active and reactive power at 30 kW and zero, respectively.

Figures 6 and 7 illustrate the dynamic behaviors at the DC side of the first and second SCES systems, respectively. In addition, their respective active and reactive powers are also plotted.

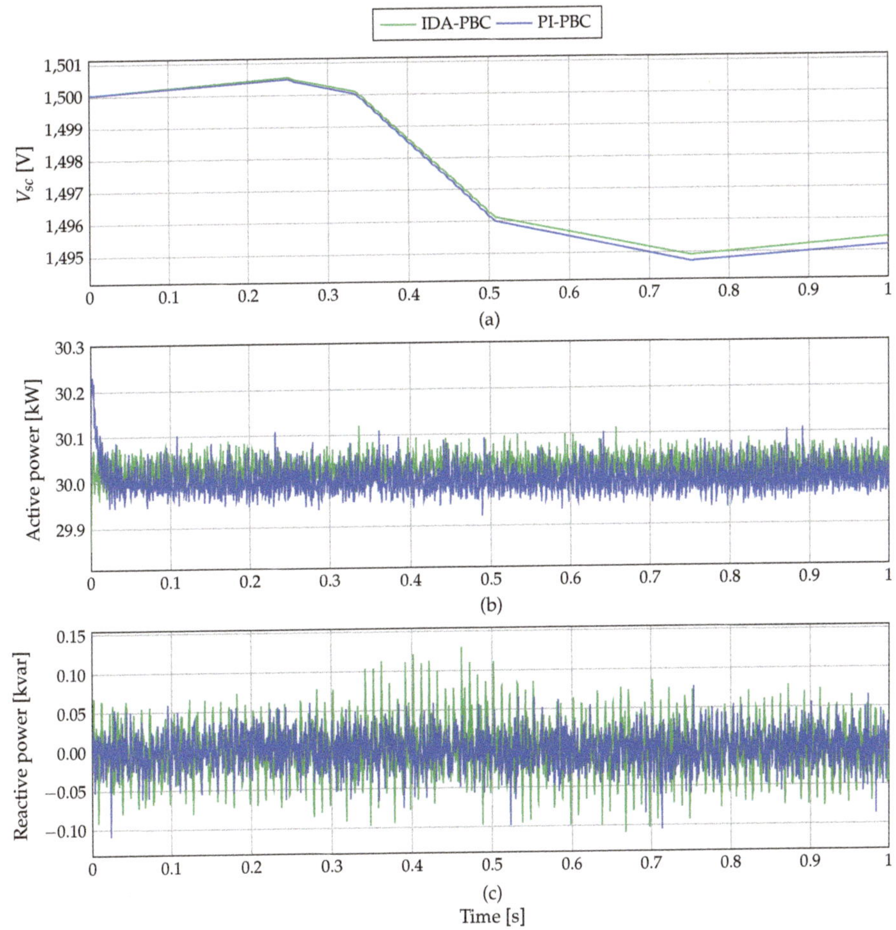

Figure 6. Dynamic behavior of the first SCES system for S2: (**a**) supercapacitor voltage; (**b**) active power provided; and (**c**) reactive power absorbed.

Observe in Figure 6 that both controllers maintain the control objectives ($p = 30$ kW and $q = 0$ kvar). However, the PI-PBC method continues presenting a better performance with lower ripples for the active and reactive power. Moreover, IDA-PBC has a steady-state error for active power of around 25 kW.

Note in Figure 7 that the proposed controller continues to show an enhanced response of active power, without steady-state error, as presented with the IDA-PBC. This behavior occurs because the IDA-PBC works as a proportional control. In contrast, the proposed controller includes an integral action that removes this error.

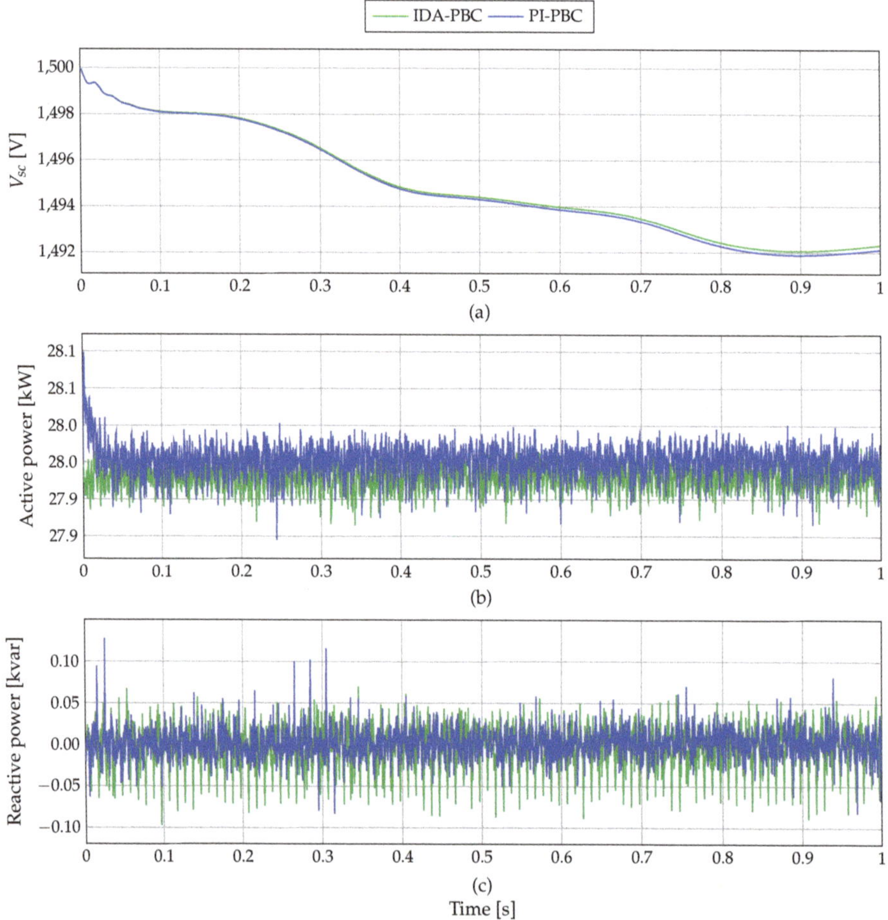

Figure 7. Dynamic behavior of the second SCES system for S2: (**a**) supercapacitor voltage; (**b**) active power provided; and (**c**) reactive power absorbed.

Complementary Analysis

Mean absolute error (MAE) and integral of time multiply absolute error (ITAE) (for the active and reactive power), and total harmonic distortion (THD) (for the AC currents) are used to quantify the performance of the controllers. Table 1 depicts these indexes for each scenario analyzed.

In Table 1, both controllers have a notable low MAE and ITAE. Nevertheless, the PI-PBC approach performs better for tracking power references than the IDA-PBC approach. This finding is supported by the reduction of MAE_p and MAE_q by 12.1% and 21.8% in the worst-case scenario (see first and the second column in Table 1), respectively. While $ITAE_p$ and $ITAE_q$ were reduced by 16.9% and 21.25% in the worst-case scenario, respectively.

Table 1. Performance Indexes.

	MAE$_p$ [W]	MAE$_q$ [var]	ITAE$_p$	ITAE$_q$	THD [%]
Scenario 1					
IDA-PBC	43.39	30.93	10.69	7.72	1.57
PI-PBC	23.71	20.65	5.94	5.35	1.55
Scenario 2 with SCES$_1$					
IDA-PBC	28.26	23.79	7.01	5.98	0.99
PI-PBC	24.68	18.59	5.82	4.64	0.98
Scenario 2 with SCES$_2$					
IDA-PBC	21.39	18.76	5.28	4.66	1.82
PI-PBC	18.38	14.60	4.36	3.63	1.80

In Table 1, it can be observed that both controllers meet the THD limits for power electronic converters established in Standard IEEE-1547 [21] even though the proposed controller has lower THD than the IDA-PBC approach. This entails that the PI-PBC approach presents better wave quality and lower losses.

The robustness of the proposed controller is investigated by applying a sensitive analysis for filter parameters. We assume that there are RL-filter mismatches with a variation of $\pm 50\%$ and $\pm 40\%$ for R and L parameters, respectively, when active power must be kept in 10 kW. For this test, we generate 100 random mismatches with uniform distribution. Figure 8 depicts the error mean value of the active power ΔP_{mean} against plant-model mismatches. Observe in this figure, that the IDA-PBC approach has a greater variation for ΔP_{mean} than the PI-PBC approach. This demonstrates that the proposed controller presents a better performance when there is a plant-model mismatch. In Figure 8, it can also be noted that the resistance variations do not influence over the controllers performance, while inductance variations do affect the performance of the controllers. This effect tends to be linear and is greater when the IDA-PBC approach is implemented.

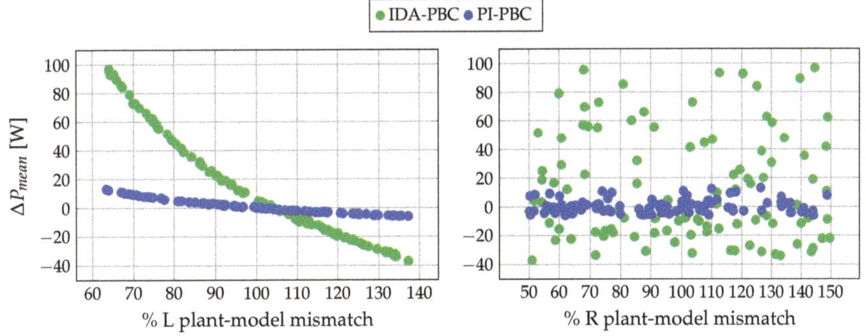

Figure 8. Sensitive analysis.

Remark 6. *The proposed controller does not show a remarkable difference in performance according to the IDA-PBC approach, which is easy to implement. This is because it only needs proportional-integral action and does not depend on the system parameters, while the IDA-PBC approach requires the system parameters and derivative calculation from the references to be applied.*

7. Conclusions

A direct power model developed from Park's reference frame for the SCES systems has been presented in this study. The direct power model controls instantaneous active and reactive powers regardless of the SCES system without employing typical phase-Locked-Loop. A PI-PBC approach was used to control the SCES system because it takes advantage of the pH structure of the SCES system to propose a control law that guarantees the stability of the system and exploits proportional-integral actions. The proposed controller demonstrates better performance in relieving the active and reactive power oscillations generated by wind generation considering imbalance when compared to the IDA-PBC approach.

Author Contributions: Conceptualization and writing—review and editing, W.G.-G. and O.D.M.; and supervision, writing—review and editing, F.M.S., C.A.R. and C.O.-H. All authors have read and agreed to the published version of the manuscript.

Acknowledgments: This work was partially supported by the National Scholarship Program Doctorates of the Administrative Department of Science, Technology and Innovation of Colombia (COLCIENCIAS), by calling contest 727-2015.

Conflicts of Interest: The authors declare no conflict of interest.

References

1. Mensah-Darkwa, K.; Zequine, C.; Kahol, P.K.; Gupta, R.K. Supercapacitor energy storage device using biowastes: A sustainable approach to green energy. *Sustainability* **2019**, *11*, 414. [CrossRef]
2. Gil, W.; Montoya, O.D.; Garces, A. Direct power control of electrical energy storage systems: A passivity-based PI approach. *Electr. Power Syst. Res.* **2019**, *175*, 105885.
3. Montoya, O.D.; Garces, A.; Espinosa-Perez, G. A generalized passivity-based control approach for power compensation in distribution systems using electrical energy storage systems. *J. Energy Storage* **2018**, *16*, 259–268. [CrossRef]
4. Aly, M.M.; Abdel-Akher, M.; Said, S.M.; Senjyu, T. A developed control strategy for mitigating wind power generation transients using superconducting magnetic energy storage with reactive power support. *Int. J. Electr. Power Energy Syst.* **2016**, *83*, 485–494. [CrossRef]
5. Montoya, O.D.; Gil-González, W.; Garces, A. SCES Integration in Power Grids: A PBC Approach under abc, $\alpha\beta0$ and dq0 Reference Frames. In Proceedings of the 2018 IEEE PES Transmission & Distribution Conference and Exhibition-Latin America (T&D-LA), Lima, Peru, 18–21 September 2018; pp. 1–5.
6. Haihua, Z.; Khambadkone, A.M. Hybrid modulation for dual active bridge bi-directional converter with extended power range for ultracapacitor application. In Proceedings of the 2008 IEEE Industry Applications Society Annual Meeting, Edmonton, AB, Canada, 5–9 October 2008; pp. 1–8.
7. Gil-González, W.J.; Garcés, A.; Escobar, A. A generalized model and control for supermagnetic and supercapacitor energy storage. *Ing. Cienc.* **2017**, *13*, 147–171. [CrossRef]
8. Thounthong, P.; Luksanasakul, A.; Koseeyaporn, P.; Davat, B. Intelligent model-based control of a standalone photovoltaic/fuel cell power plant with supercapacitor energy storage. *IEEE Trans. Sustain. Energy* **2012**, *4*, 240–249. [CrossRef]
9. Mufti, M.D.; Iqbal, S.J.; Lone, S.A.; Ain, Q. Supervisory Adaptive Predictive Control Scheme for Supercapacitor Energy Storage System. *IEEE Syst. J.* **2015**, *9*, 1020–1030. [CrossRef]
10. Montoya, O.D.; Gil-González, W.; Garces, A. Distributed energy resources integration in single-phase microgrids: An application of IDA-PBC and PI-PBC approaches. *Int. J. Electr. Power Energy Syst.* **2019**, *112*, 221–231. [CrossRef]
11. Montoya, O.D.; Gil-González, W.; Avila-Becerril, S.; Garces, A.; Espinosa-Pérez, G. Distributed Energy Resources Integration in AC Grids: A Family of Passivity-Based Controll, (in Spanish). *Rev. Iberoam. Autom. Inform. Ind.* **2019**, *16*, 212–221. [CrossRef]
12. Ortega, R.; Perez, J.A.L.; Nicklasson, P.J.; Sira-Ramirez, H.J. *Passivity-Based Control of Euler-Lagrange Systems: Mechanical, Electrical and Electromechanical Applications*; Springer Science & Business Media: Berlin/Heidelberg, Germany, 2013.

13. van der Schaft, A. *L2-Gain and Passivity Techniques in Nonlinear Control*; Springer: Berlin/Heidelberg, Germany, 2017.

14. Cisneros, R.; Pirro, M.; Bergna, G.; Ortega, R.; Ippoliti, G.; Molinas, M. Global tracking passivity-based PI control of bilinear systems: Application to the interleaved boost and modular multilevel converters. *Control Eng. Pract.* **2015**, *43*, 109–119. [CrossRef]

15. Zonetti, D. Energy-Based Modelling and Control of Electric Power Systems with Guaranteed Stability Properties. Ph.D. Thesis, Université Paris-Saclay, Saint-Aubin, France, 2016.

16. Zonetti, D.; Ortega, R.; Benchaib, A. A globally asymptotically stable decentralized PI controller for multi-terminal high-voltage DC transmission systems. In Proceedings of the 2014 European control conference (ECC), Strasbourg, France, 24–27 June 2014; pp. 1397–1403.

17. Zonetti, D.; Ortega, R.; Benchaib, A. Modeling and control of HVDC transmission systems from theory to practice and back. *Control. Eng. Pract.* **2015**, *45*, 133–146. [CrossRef]

18. Gil-González, W.; Montoya, O.D.; Garces, A. Direct power control for VSC-HVDC systems: An application of the global tracking passivity-based PI approach. *Int. J. Electr. Power Energy Syst.* **2019**, *110*, 588–597. [CrossRef]

19. Montoya, O.D.; Gil-González, W.; Serra, F.M. PBC Approach for SMES Devices in Electric Distribution Networks. *IEEE Trans. Circuits Syst. II Exp. Briefs* **2018**, *65*, 2003–2007. [CrossRef]

20. Harnefors, L.; Nee, H.P. Model-based current control of AC machines using the internal model control method. *IEEE Trans. Ind. Appl.* **1998**, *34*, 133–141. [CrossRef]

21. IEEE Standard for Interconnecting Distributed Resources with Electric Power Systems—Amendment 1. In *IEEE Std 1547a-2014 (Amendment to IEEE Std 1547-2003)*; IEEE: Piscataway, NJ, USA, 2014; pp. 1–16. [CrossRef]

Article

Optimal Location and Sizing of PV Sources in DC Networks for Minimizing Greenhouse Emissions in Diesel Generators

Oscar Danilo Montoya [1,2], Luis Fernando Grisales-Noreña [3], Walter Gil-González [2], Gerardo Alcalá [4] and Quetzalcoatl Hernandez-Escobedo [5,*]

1 Facultad de Ingeniería, Universidad Distrital Francisco José de Caldas, Carrera 7 No. 40B - 53, Bogotá D.C 11021, Colombia; o.d.montoyagiraldo@ieee.org or omontoya@utb.edu.co
2 Laboratorio Inteligente de Energía, Universidad Tecnológica de Bolívar, Km 1 vía Turbaco, Cartagena 131001, Colombia; wjgil@utp.edu.co
3 Departamento de Electromecánica y Mecatrónica, Instituto Tecnológico Metropolitano, Medellín 050012, Colombia; luisgrisales@itm.edu.co
4 Centro de Investigación en Recursos Energéticos y Sustentables, Universidad Veracruzana, Coatzacoalcos, Veracruz 96535, Mexico; galcala@uv.mx
5 Escuela Nacional de Estudios Superiores Juriquilla, UNAM, Queretaro 76230, Mexico
* Correspondence: qhernandez@unam.mx

Received: 27 December 2019; Accepted: 10 February 2020; Published: 24 February 2020

Abstract: This paper addresses the problem of the optimal location and sizing of photovoltaic (PV) sources in direct current (DC) electrical networks considering time-varying load and renewable generation curves. To represent this problem, a mixed-integer nonlinear programming (MINLP) model is developed. The main idea of including PV sources in the DC grid is minimizing the total greenhouse emissions produced by diesel generators in isolated areas. An artificial neural network is employed for short-term forecasting to deal with uncertainties in the PV power generation. The general algebraic modeling system (GAMS) package is employed to solve the MINLP model by using the CONOPT solver that works with mixed and integer variables. Numerical results demonstrate important reductions of harmful gas emissions to the atmosphere when PV sources are optimally integrated (size and location) to the DC grid.

Keywords: artificial neural networks; diesel generation; direct current networks; greenhouse emissions; numerical optimization; mixed-integer nonlinear programming photovoltaic plants

1. Introduction

Recent deployments of power electronics have allowed the positive advancement and development of efficient renewable energy interfaces for wind and photovoltaic plants [1–3], and the large-scale usage of energy storage systems, such as batteries [4–6], supercapacitors [7,8], superconductors [3,9,10] and flywheels [11,12]. These devices can be integrated into the power system using different interface technologies; i.e., to operate in the classical alternating current (AC) or direct current (DC) grids [3,13]. The selection of the grid operative condition plays an important role in the quality of the electrical service provided to the end-users. In the case of AC grids, it is required to maintain sinusoidal voltages with adequate form (power quality criteria); i.e., magnitude, frequency, power factor, harmonics, etc. [14–16]. These characteristics in AC networks make them more complex in comparison to DC networks, since these latter only require voltage control, and reactive power and frequency are nonexistent [17,18]. An additional advantage of using DC over AC technologies is their high efficiency in terms of power loss and voltage profiles [19]. These features can be added to the fact that photovoltaic plants or some energy storage technologies (supercapacitors, batteries and superconducting coils) work directly in the DC paradigm [17,20], which can help to reduce the number of power interfaces to integrate these technologies in DC grids in comparison to their AC counterparts [21].

In specialized literature, DC networks have taken relevance regarding optimization and control applications [21]. In the case of optimization, multiple approaches based on semidefinite programming [4], second-order cone programming [20], sequential quadratics [22] and metaheuristics have been proposed to address optimal power flow problems [23]; additionally, some classical numerical methods can be found, such as Newton-Raphson [24], Gauss-Seidel [17] or successive approximations for power flow solutions [25]. In the case of control, the most conventional approaches focus on battery control [26], renewable energy integration [27] and dynamic stability based on passivity based-control [28,29], model predictive control [30,31] and sliding mode control [32,33].

These recent studies show that DC networks are promissory technologies that require extensive research for being successfully operated and also integrated and interconnected to the conventional AC power system [34,35]. In this paper, we deal with a classical and well-studied problem of optimal location and sizing of distributed generation in power systems. Nevertheless, we focus on direct current networks in isolated areas operated with diesel generators [36,37]. Although this problem has been widely studied in AC networks with metaheuristics, such as genetic algorithms [38], tabu search [39], harmonic search [40], krill-herd algorithm [41] and population based-learning methods [42], in conjunction with exact mixed-integer nonlinear programming methods [43,44], on the topic of DC networks there are only four references that address this problem. In [45] a semidefinite programming method was proposed for binary variable relaxation associated with the location of the generators; then, its binary structure was recovered with random hyperplanes; in [46] a sequential quadratic programming model with the same binary relaxation was proposed, and the binary nature of the problem was recovered with a heuristic search that defines the optimal location of the generators. The authors of [47] have addressed the issue of optimal location and sizing of distributed generators in DC networks from the metaheuristic point of view by using a classical genetic algorithm for their locations, in conjunction with their different optimal power flow methods based on particle swarm, black hole and continuous genetic optimizers. In [48], a mixed-integer nonlinear programming model was proposed to locate and size DGs in DC microgrids, using the general algebraic modeling system (GAMS) package for its solution. The common denominator of these approaches is the fact that load or renewable generation variations are not considered, since all of them only solve the problem for a unique hour time lapse. This cannot replicate the real behavior of electrical networks, especially when renewable energy resources are introduced.

To deal with renewable energy variations in the problem of optimal location and sizing of distributed generators in DC networks, here, we propose mixed-integer nonlinear programming for location and sizing of PV plants in DC networks to minimize the total pollution and greenhouse emissions released by diesel generators feeding isolated DC networks. The main contribution of our approach is based on a multiperiod optimization problem, including expected curves of PV generation and demand consumption focused on Colombian power system characteristics. To ensure that PV generation potential was well estimated, an artificial neural network with recursive connections was employed. To solve the proposed mixed-integer nonlinear programming (MINLP) model, we used the GAMS package with multiple nonlinear solvers to compare the results, as recommended in [44,48]. Numerical results confirm that the correct placement of PV plants helps via a significant reduction of pollutants released to the atmosphere by fossil fuels, which contributes positively to the responsible plans of energy consumption for future generations; i.e., making the power system more sustainable [6].

The remainder of this document is organized as follows: In Section 2 is presented the MINLP model for the problem of optimal location and sizing of PV generators in DC networks, for the reduction of greenhouse emissions with multiperiod structure. In Section 3 it the artificial neural network employed to forecast the solar power generation is described. In Section 4 a GAMS example in a small DC network to solve the problem addressed in this paper is presented. In Section 5 the numerical simulation in a 21-nodes test feeder is shown to minimize pollutants produced by diesel generators with its corresponding analysis and discussion. In section 6 the main concluding remarks derived from this research are presented.

2. Mathematical Model

The problem of the optimal location of PV plants in DC networks considering load variations corresponds to a nonlinear, non-convex and non-differentiable optimization problem that combines discrete and continuous variables, which generates an MINLP problem [47]. The main interest in this formulation is to minimize the greenhouse emissions produced by diesel generators interconnected to DC rural (isolated) networks [48,49]. The complete mathematical model is presented below:

Objective function:

$$\min z = \sum_{t=1}^{T} \sum_{i=1}^{N} R_i^{ge} p_{i,t}^{cg} \Delta t, \tag{1}$$

where z are the total greenhouse emissions in pounds; R_i^{ge} is the rate of greenhouse emissions per kilowatt-hour; $p_{i,t}^{cg}$ is the total power generation in the conventional generators (diesel sources); and Δt is the period of time under analysis, typically $\Delta t = 1$ h. Note that N is the total number of nodes in the DC grid and T is the number of periods of time, N and T being the sizes of the sets of nodes $\mathcal{N}$ and periods of time $\mathcal{T}$, respectively.

Set of constraints:

$$p_{i,t}^{cg} + y_i^{pv} p_{i,t}^{pv,nom} - p_{i,t}^{d} = v_{i,t} \sum_{j=1}^{N} G_{ij} v_{j,t}, \qquad \forall \{i \in \mathcal{N}, t \in \mathcal{T}\} \tag{2}$$

$$i_{ij,t} = g_{ij} \left(v_{i,t} - v_{j,t} \right), \qquad \forall \{ij \in \mathcal{L}, t \in \mathcal{T}\} \tag{3}$$

$$p_{i,t}^{cg,min} \leq p_{i,t}^{cg} \leq p_{i,t}^{cg,max}, \qquad \forall \{i \in \mathcal{N}, t \in \mathcal{T}\} \tag{4}$$

$$-i_{ij}^{max} \leq i_{ij,t} \leq i_{ij}^{max}, \qquad \forall \{ij \in \mathcal{L}, t \in \mathcal{T}\} \tag{5}$$

$$v_i^{min} \leq v_{i,t} \leq v_i^{max}, \qquad \forall \{i \in \mathcal{N}, t \in \mathcal{T}\} \tag{6}$$

$$0 \leq y_i^{pv} \leq p_i^{pv,max} x_i^{pv}, \qquad \forall \{i \in \mathcal{N},\} \tag{7}$$

$$\sum_{i=1}^{N} x_i^{pv} \leq NG_{pv}^{max}, \tag{8}$$

where $p_{i,t}^{pv,nom}$ and $p_{i,t}^{d}$ are the nominal power injection of the PV generator and the power consumption in the node i during the period of time t; y_i^{pv} defines the size of the PV generator connected to the node i; $p_{i,t}^{pv,nom}$ is the nominal power of the PV generators, which is dependent on the solar forecasting in the zone of influence of the DC network; $v_{i,t}$ and $v_{j,t}$ represent the voltage value in the nodes i and j respectively during the time t; G_{ij} is the component of the conductance matrix that relates nodes i and j, whose value depends on the physical connections between nodes (i.e., it is dependent on the grid configuration); $i_{ij,t}$ represents the value of the current that flows between nodes i and j in the period time t, which depends on the conductor conductance named g_{ij}; $p_{i,t}^{cg,min}$ and $p_{i,t}^{cg,max}$ represent the minimum and maximum capabilities of power generation in the diesel generators connected at the node i in the period of time t; $p_i^{pv,max}$ is the maximum nominal size of the PV source that can be connected in the node i; i_{ij}^{max} corresponds to the maximum current that can flow through the conductor that connects nodes i and j; v_i^{min} and v_i^{max} are the minimum and maximum voltage bounds allowed at each node; x_i^{pv} is a binary variable that defines whether a PV source is installed ($x_i^{pv} = 1$ if the PV source is installed and $x_i^{pv} = 0$ otherwise) at node i; NG_{pv}^{max} defines the maximum number of PV generators available to be located into the grid. (The components G_{ij} and g_{ij} have the same magnitude;

notwithstanding, these differ by sign, $g_{ij} = \frac{1}{r_{ij}}$ being positive; i.e., $G_{ij} = -g_{ij}$. r_{ij} is the resistance value of the conductor located between nodes i and j.)

The mathematical model defined from (1) to (8) has the following interpretation [22,50]: the objective function (1) quantifies the total greenhouse emissions produced by the diesel generators during their operation; in (2) is presented the power balance equation per node, which is typically known as power flow constraint, this being a non-affine constraint. Expression (3) shows the calculation of the current flow through ijth branch as a function of the voltage drop; in (4) is defined the box constraint related with the power capabilities in the diesel generators; in (5) the thermal bounds of the network conductors are presented in Amperes; meanwhile, (6) defines the voltage regulation bounds of the grid, which are defined by regulatory entities. Expression (7) determines the possibility of locating a PV generator in the network by defining its maximum ranks admissible for a generation. Finally, in (8) is presented the constraint associated with the maximum number of generators available for installation.

Note that the mathematical optimization model is complex to be solved, since it combines binary and integer variables, this model being an MINLP [45]. Additionally, the most complicated constraint is the power balance defined in (2), since it represents the hyperbolic relation between voltage and currents in electrical DC networks with constant power loads [51], which is a non-affine nonlinear constraint without convex properties.

Even though in specialized literature it has been reported, some approaches to solving this problem by decoupling it into a master-slave optimization problem with metaheuristics (i.e., genetic algorithms with particle swarm derived approaches [47]), no reports with the multiperiod structure (1)–(8) were found. For this reason, we are concentrating on the mathematical formulation and not on the solution technique. Since recent publications use the GAMS package and its large-scale nonlinear optimizers to solve similar problems [44,48], we decided to use the GAMS package in conjunction with the CONOPT solver as a solution strategy.

3. Solar Generation Forecasting

Planification of electrical networks that include renewable generators is a challenge, since it is necessary to consider their intermittency into the planning project. Due to that, in this research, we are interested in locating and sizing PV plants in DC networks to diminish greenhouse emissions produced by diesel generators for isolated areas, it being strictly necessary to know the PV potential in these areas. For this purpose, we consider that this electrical network is located in the Caribbean region in Colombia, and the solar power availability measured during a year is presented in Figure 1.

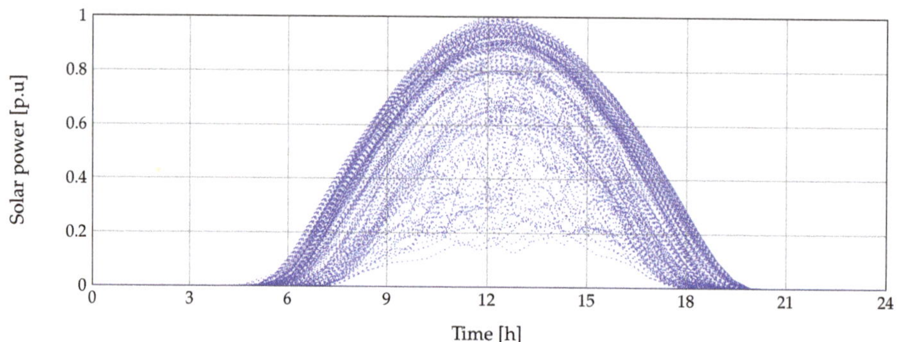

Figure 1. Historical data for solar prediction (adapted from [4,50]).

On the other hand, the uncertainty of primary sources for PV plants due to temperature and solar radiation produces a challenge for their optimal location and sizing. An improper location or sizing of PV plants can generate problems in the voltage profiles or cause transmission line overload and increase power loss [4]. Therefore, it is necessary to consider a methodology that takes into account the high variability of the temperature and solar radiation with the purpose of reducing the forecasting errors, and thus, avoiding problems that can be introduced due to incorrect location or sizing. For this purpose, we employ the methodology developed in [50], which uses an artificial neural network (ANN) in order to estimate the primary sources for PV plants adequately.

Artificial Neural Network

Artificial neural networks have been largely used to solve multiple problems in engineering, and are based on artificial intelligence [4]. They have a wide range of applications, from pattern classification and clustering, to optimization and prediction, among others [50]. Here the ANNs are used to predict the most probable temperature and solar radiation.

Typically, the ANNs go through three processes: training, adjusting and validating. There are multiples nonlinear learning rules to use in the training. We used the following rule to train the ANN:

$$y(t) = f\left(y(t-1), ..., y(t-n_y), x(t-1), ..., x(t-n_x)\right) \tag{9}$$

where x and y are input and output data, respectively. n_y and n_x are the last values of the prediction and the input data, respectively.

The ANN applied to solar radiation forecasting has been trained, adjusted and validated with 70%, 15% and 15% of the data, respectively. Additionally, two inputs (time and temperature), six delays and 18 hidden neurons have been implemented on the ANN. This was implemented in MATLAB R2019b employing *ntstool*. Figure 2 depicts its schematic implementation.

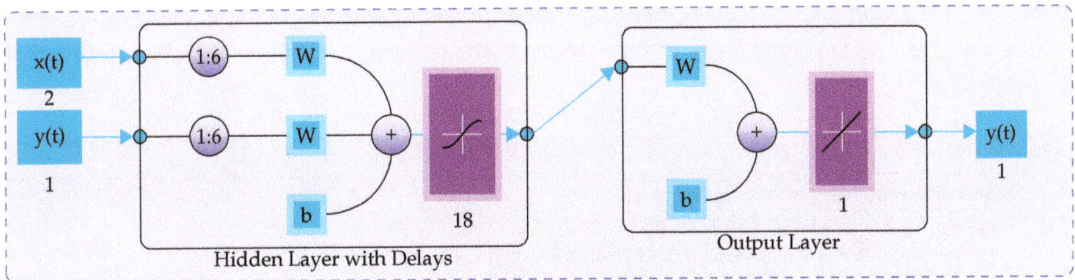

Figure 2. ANN scheme for solar radiation prediction [50].

The ANN scheme illustrated in Figure 2 contains a two-layer feedforward network. The first is the hidden layer that works with a sigmoid transfer function, while the second one is the output layer, which consists of a linear transfer function. The hidden layer uses delays to stores previous values of input $x(t)$ and output $y(t)$ data. W and b are weights and bias values in the training process of the proposed ANN, respectively. The output $y(t)$ in the output layer also applies rectified linear unit (ReLU) layer, which consists of

$$f(y) = \begin{cases} y, & y \geq 0, \\ 0, & y < 0. \end{cases} \tag{10}$$

The ReLU layer is used since in some cases the radiation estimation may be negative.

In the training process of the proposed ANN for solar forecasting the Levenberg–Marquardt algorithm was employed and it is available for the *ntstool* in MATLAB [52]. The main advantage of this training process in relation to classical Newton algorithm or Gauss-Newton algorithm is the rate of convergence and its stability. More information about the Levenberg-Marquardt algorithm can be found in [53,54].

4. Optimization Strategy

The solution of the mathematical model described from (1) to (8) requires a methodology that works with mixed-integer variables [45]. In the specialized literature, two main approaches have been proposed to deal with MINLP models in power systems. One of them corresponds to the hybridization methods composed by master-slave stages with metaheuristics. Some of them are genetic algorithms [38], the particle swarm optimizer [42], tabu search algorithms [39] and krill-herd algorithms [41]. These methods decouple the problem of PV source location from sizing. The solutions of these problems are reached through iterative procedures; nevertheless, the optimal solution depends on the number of iterations defined, with the main disadvantage that it is not possible to find the same numerical solution each time that the methodology is evaluated [42]. Thus, statistical procedures are needed to determine their efficiencies [46]. The second focus is to solve MINLP models using gradient-based approaches embedded into branch and bound (B&B) methods, where the gradient searches solve the resulting nonlinear programming model, while the B&B guides the discrete search by defining the location of the PVs [43]. These approaches use nonlinear large-scale solvers available in optimization packages such as GAMS [6,44]. Here, due to the contribution of this paper being related to the presentation of the MINLP model to locate and size PV sources in DC grids for isolated areas to reduce greenhouse emissions by diesel generators, we solve the proposed mathematical model using the CONOPT solver available for GAMS.

As mentioned before, this solver works with gradient searches and B&B methods. In the first step, all the binary variables are relaxed to find the best possible solution; then, this relaxation is discretized to recover the nature of the binary variables in order to provide the optimal solution of the problem. It is important to mention that this optimization package has been successfully used in different problems, such as the optimal operation of batteries [6], optimal location of distributed generators in AC and DC grids [44], optimal design of osmotic power plants [55] and economic dispatch analysis [50,56]. Finally, Algorithm 1 resumes the necessary steps to solve the MINLP model defined from (1) to (8) [57].

Algorithm 1: Main steps for solving the proposed MINLP model in GAMS [57]

    Select the test system characteristics;
    Define the sets involved in the optimization model;
    Define the scalars, parameters and matrices of the DC grid;
    Define the variables and their bounds;
    Define the name of the equations and build the mathematical model (1) to (8);
    Determine the number of PV to be located;
    Select the CONOPT solver in GAMS;
    Define the variables of interest to be visualized;
    Read and analyze the final results;

---

5. Test System and Numerical Validations

In this section, we present the test system structure and the output behavior of the PV forecasting used as input on the location and sizing process; in addition, all the simulation scenarios are defined and the numerical results are presented using the GAMS optimization package with its large-scale nonlinear solver CONOPT [44].

5.1. Test System

As the DC distribution system, we consider a 21-node test system with two slack (diesel) generators located at nodes 1 and 21, which support voltage profiles in the grid with 1.0 and 1.05 p.u, respectively. The configuration of this test system is depicted in Figure 3 and the base values of this test system are 1 kV and 100 kW [4].

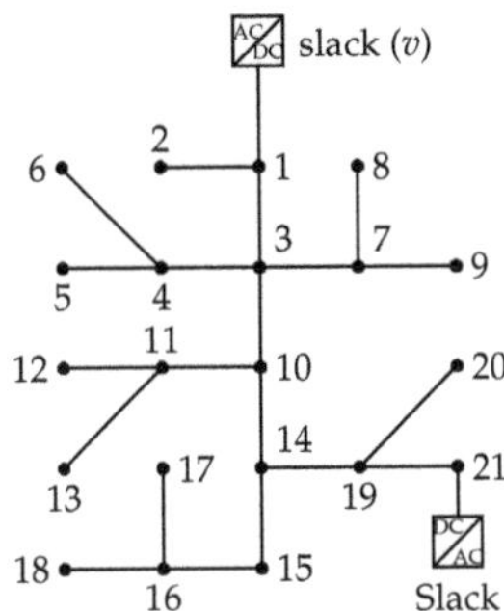

Figure 3. Electrical configuration for the 21-node test system.

Table 1 reports the numerical information of this test system, where its interpretation from left-to-right is as follows: Sending node, receiving node, branch resistance and power consumption at the receiving node. Note that in the case of the diesel generator located at node 21, there is a constant power consumption connected to its node.

Note that the total power consumption of this test feeder at the peak load is 554 kW.

Table 1. Electrical parameters of the 21-node test system.

Node i	Node j	R_{ij} [pu]	P_j [pu]	Node i	Node j	R_{ij} [pu]	P_j [pu]
1(slack)	2	0.0053	0.70	11	12	0.0079	0.68
1	3	0.0054	0.00	11	13	0.0078	0.10
3	4	0.0054	0.36	10	14	0.0083	0.00
4	5	0.0063	0.04	14	15	0.0065	0.22
4	6	0.0051	0.36	15	16	0.0064	0.23
3	7	0.0037	0.00	16	17	0.0074	0.43
7	8	0.0079	0.32	16	18	0.0081	0.34
7	9	0.0072	0.80	14	19	0.0078	0.09
3	10	0.0053	0.00	19	20	0.0084	0.21
10	11	0.0038	0.45	19	21(slack)	0.0082	-0.21

5.2. Objective Function and Daily Curves

To determine the number of greenhouse emissions by using diesel generators, we consider the information reported in [6]. In this reference the number of different greenhouse emissions caused by diesel generators with sizes smaller than 1 MW is presented. This information is reported in Table 2.

Table 2. Main gasses released by diesel generators with capacities lower than 1 MW.

Type of Emission	Chemical Symbol	Rank [lb/MWh]
Carbon dioxide	CO_2	1000–1700
Sulfur dioxide	SO_2	0.40–3.00
Nitrogen oxides	NO_x	10–41
Carbon monoxide	CO	0.40–9.00
Heavy particles	$PM-10$	0.40–3.00

Due to the most important emissions being carbon dioxide (CO_2), we select the average value in the rank, i.e., 1350 lb/MWh, to consider in the objective function. In the case of renewable generation prediction, in Table 3 a typical curve is reported for a sunny day in the Caribbean region in Colombia, as is the prediction reached by the proposed ANN. (This information is presented numerically to guarantee that results can be reproduced in future research.) In addition, we include a curve for power consumption provided by a utility in Colombia. (The name of the utility is not provided due to confidential agreements.) Note that these curves are normalized to make them independent of the size of the PV source or the size of the utility [44,50].

Table 3. Real and predictive PV curves and load behavior during a typical sunny day in Colombia.

Period	Real [pu]	Forec. [pu]	Load [pu]	Period	Real [pu]	Forec. [pu]	Load [pu]
1	0.000	0.000	0.633	25	1.000	0.976	0.814
2	0.000	0.000	0.619	26	0.975	1.000	0.842
3	0.000	0.000	0.605	27	0.771	0.978	0.869
4	0.000	0.000	0.578	28	0.889	0.790	0.886
5	0.000	0.000	0.550	29	0.630	0.883	0.902
6	0.000	0.000	0.495	30	0.593	0.604	0.905
7	0.000	0.000	0.440	31	0.404	0.606	0.908
8	0.000	0.000	0.435	32	0.366	0.357	0.908
9	0.000	0.000	0.429	33	0.231	0.328	0.908
10	0.000	0.000	0.421	34	0.203	0.142	0.935
11	0.000	0.000	0.413	35	0.130	0.142	0.963
12	0.000	0.000	0.419	36	0.053	0.073	0.987
13	0.000	0.000	0.426	37	0.008	0.019	0.988
14	0.000	0.000	0.433	38	0.000	0.008	0.989
15	0.000	0.026	0.440	39	0.000	0.000	0.990
16	0.024	0.052	0.495	40	0.000	0.000	0.995
17	0.124	0.110	0.550	41	0.000	0.000	1.000
18	0.272	0.263	0.550	42	0.000	0.000	0.995
19	0.439	0.431	0.550	43	0.000	0.000	0.990
20	0.604	0.594	0.605	44	0.000	0.000	0.935
21	0.733	0.730	0.660	45	0.000	0.000	0.880
22	0.810	0.830	0.701	46	0.000	0.000	0.770
23	0.860	0.875	0.743	47	0.000	0.000	0.660
24	0.984	0.899	0.778	48	0.000	0.000	0.633

5.3. Simulation Scenarios

To evaluate our proposed mathematical model for optimal location and sizing of PV sources in DC grids, we consider that the size of these generators can be 60% of the total demand in the peak hour; i.e., 332.4 kW. In addition, we consider four simulation cases as follows: $\mathbf{S}_1$) as the base case of the test system without including PV sources, $\mathbf{S}_2$) as the location of one PV source, $\mathbf{S}_3$) as the location of two PV sources and $\mathbf{S}_4$) as the location of three PV sources inside of the DC system. The main idea of these scenarios is to evidence the effect of having a total PV installation capacity distributed in different quantities of injection depending on the simulation case.

5.4. Numerical Results

To solve the general MINLP model, which represents the problem of optimal location and sizing of DGs in DC systems, we employed the GAMS optimization package with different nonlinear solvers in a desktop computer with an INTEL(R) Core(TM) i5-3550 3.5-GHz processor and 8 GB of RAM running a 64-bit version of Windows 10 Home Single Language.

Table 4 reports the numerical results for all the simulation scenarios previously proposed. Note that in the first scenario, the daily greenhouse emissions of CO_2 are 13,428.912 pounds, and the minimum emissions occur in the fourth scenario with 10,878.190 pounds per day, which implies en equivalent reduction of 2550.722 pounds of emissions of CO_2 per day (18.99%). It is important to mention that after using the CONOPT solver for each simulation scenario, the total processing time to reach the optimal solution is lower than 20 s for all the cases. Still, it starts to increase from 6.224 s to 19.063 s (67.35%) depending on the number of PV sources that are considered in the scenario. It is important to mention that the processing times of the ANN training process is not considered in the last column in Table 4, since this is an offline procedure that takes between 5 and 10 min depending on the size of the training set.

Table 4. Gas emissions for each simulation case.

Simulation Scenario	Objective Function [lb] (CO_2)	Processing Time [s]
S_1	13,428.91	6.224
S_2	11,027.19	11.001
S_3	10,892.80	18.478
S_4	10,878.18	19.063

Figure 4 presents the total daily reduction per day when a different number of PV sources are located and sized in the 21 nodes DC test feeder. This plot confirms that the best scenario corresponds to the case with three PV sources located inside the network; nevertheless, these bars confirm that the reduction of greenhouse emissions has a nonlinear behavior regarding the number of PV sources; i.e., the reduction tends to have a saturation of 19% approximately. This behavior obeys the fact that the PV sources only work during sunny hours, making it necessary to use diesel sources during the night.

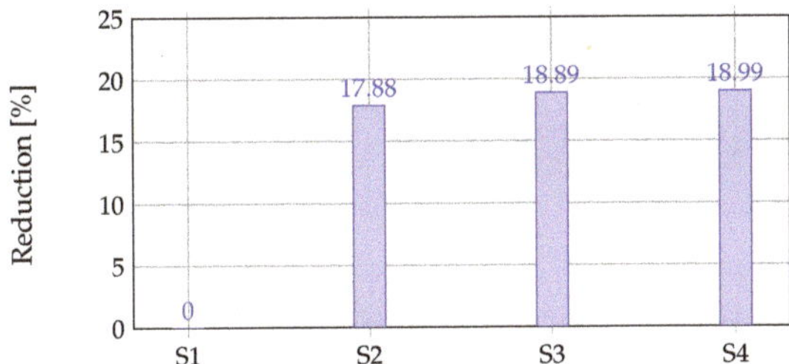

Figure 4. Total reductions in CO_2 emissions per day for each simulation scenario.

Table 5 reports the optimal location of the PV sources in each simulation scenario. These simulations show that the most attractive node to locate a PV source is the node 17, followed by nodes 19 and 12, respectively. In addition, for the third and fourth scenarios, it is important to observe that all the allowed penetration, i.e., 60% of the peak consumption, is used (divided) by the PV generators, while the second scenario only uses about 96.38% of this maximum capability. These results confirm the nonlinear relation between the number of sources available for location and their sizes regarding the minimization of the objective function, which implies that multiple simulations and scenarios need to be taken into account for the grid planner (i.e., utility) to determine its inversions.

Table 5. Optimal locations and sizes of the PV sources.

Simulation Scenario	Location [node]			Size [kW]			Total Penetration [kW]
S_1	—	—	—	—	—	—	0
S_2	17	—	—	320.37	—	—	320.372
S_3	17	19	—	141.30	191.098	—	332.400
S_4	12	17	19	91.33	101.582	139.485	332.400

Note that all the simulations were guaranteed through (5) to have all the currents flowing inside of the system be lower than 400 A; i.e., all the possible locations and sizes of the PV sources are feasible to be implemented since conductors of the grid can operate safely. In addition, when we considered the real daily curve reported in Table 3 with the locations and sizes of the PV sources reported in Table 5, it was found that the errors in the estimation of the objective function were lower than 1%, which confirms that ANN are powerful tools for short-term forecasting of renewable energy resources, as reported in [50].

Figure 5 presents the behavior of the diesel generators during the day for all the simulated scenarios. Note that in the first scenario, they support all the power consumption in the DC grid (solid line in Figure 5a,b). Nevertheless, when PV generators are installed, the total generation in the diesel resources decreases significantly between hours 7 and 18; which clearly corresponds to the periods of time where PV sources can deliver their power to the grid (see dotted and dashed lines in Figure 5a,b).

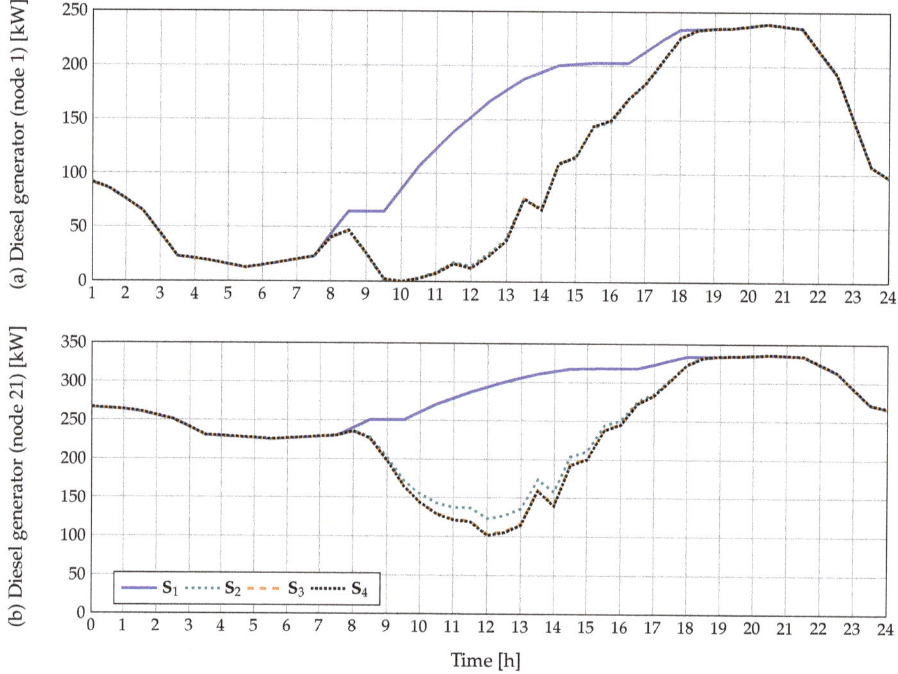

Figure 5. Generation profiles of the diesel sources during the day.

It is also important to mention that the differences between both profiles in the diesel generators are caused by the voltage profiles required at their terminals. In the case of the diesel generator located at node 21, it needs to inject more active power, since it is required to maintain the voltage at 1.05 p.u all the time, while the diesel generator located at node 1 was fixed to 1.0 p.u. This implies that with lower power injections this profile can be sustained.

6. Conclusions and Future Works

A mixed-integer nonlinear programming model for optimal location and sizing of PV sources in DC isolated networks was presented in this paper. The objective of this formulation is to reduce the total greenhouse emissions (i.e., pounds of CO_2 emissions per day) by diesel generators with the introduction of PV sources considering typical Colombian power profiles and consumption behaviors. Numerical results demonstrate that these gas emissions can be reduced between 17% and 19% depending on the number of PV sources installed and their sizes.

A nonlinear relation between the number of generators and their location was evidenced with the minimization of the objective function, as differences lower than 1.50% were found for all the scenarios that included PV sources (i.e., from the second to the fourth scenario). These situations imply that additional studies are needed in regard to operational costs, useful life and ground lot availability, among other things, to determine the best solution in terms of PV sizes and locations. The objective function, in all scenarios provided in this studio, showed attractive alternatives to be implemented in the case of the 21-node test feeder.

As future work, it would be possible to extend this MINLP model to wind generators by predicting their average power availability with artificial neural networks with high efficiency, as reported in this study of PV sources. In addition, this model can be modified to include battery energy storage systems, to increase the introduction of renewable energy during periods of time with high demand and lower generation availability.

Author Contributions: Conceptualization, O.D.M., L.F.G.-N., W.G.-G., G.A. and Q.H.-E.; methodology, O.D.M., L.F.G.-N., W.G.-G., G.A. and Q.H.-E.; formal analysis, O.D.M., L.F.G.-N., W.G.-G., G.A. and Q.H.-E.; investigation, O.D.M., L.F.G.-N., W.G.-G., G.A. and Q.H.-E.; resources, O.D.M., L.F.G.-N., W.G.-G., G.A. and Q.H.-E.; writing—original draft preparation, O.D.M., L.F.G.-N., W.G.-G., G.A. and Q.H.-E. All authors have read and agreed to the published version of the manuscript.

Funding: This research was funded by the Universidad Tecnológica de Bolívar under project CP2019P011, and Instituto Tecnológico Metropolitano, Universidad Nacional de Colombia, and Colciencias under the project "Estrategia de transformación del sector energético Colombiano en el horizonte de 2030 - Energética 2030"—"Generación distribuida de energía eléctrica en Colombia a partir de energía solar y eólica" (Code: 58838, Hermes: 38945).

Conflicts of Interest: The authors declare no conflict of interest.

References

1. Blaabjerg, F.; Chen, Z.; Kjaer, S.B. Power electronics as efficient interface in dispersed power generation systems. *IEEE Trans. Power Electron.* **2004**, *19*, 1184–1194. [CrossRef]

2. Wandhare, R.G.; Agarwal, V. Novel Integration of a PV-Wind Energy System With Enhanced Efficiency. *IEEE Trans. Power Electron.* **2015**, *30*, 3638–3649. [CrossRef]

3. Montoya, O.D.; Gil-González, W.; Garces, A. Distributed energy resources integration in single-phase microgrids: An application of IDA-PBC and PI-PBC approaches. *Int. J. Electr. Power Energy Syst.* **2019**, *112*, 221–231. [CrossRef]

4. Gil-González, W.; Montoya, O.D.; Holguín, E.; Garces, A.; Grisales-Noreña, L.F. Economic dispatch of energy storage systems in dc microgrids employing a semidefinite programming model. *J. Energy Storage* **2019**, *21*, 1–8. [CrossRef]

5. Grisales, L.F.; Montoya, O.D.; Grajales, A.; Hincapie, R.A.; Granada, M. Optimal Planning and Operation of Distribution Systems Considering Distributed Energy Resources and Automatic Reclosers. *IEEE Lat. Am. Trans.* **2018**, *16*, 126–134. [CrossRef]

6. Montoya, O.D.; Grajales, A.; Garces, A.; Castro, C.A. Distribution Systems Operation Considering Energy Storage Devices and Distributed Generation. *IEEE Lat. Am. Trans.* **2017**, *15*, 890–900. [CrossRef]

7. Di Noia, L.P.; Genduso, F.; Miceli, R.; Rizzo, R. Optimal Integration of Hybrid Supercapacitor and IPT System for a Free-Catenary Tramway. *IEEE Trans. Ind. Appl.* **2019**, *55*, 794–801. [CrossRef]

8. Jayasinghe, S.D.G.; Vilathgamuwa, D.M.; Madawala, U.K. A Dual Inverter-Based Supercapacitor Direct Integration Scheme for Wind Energy Conversion Systems. *IEEE Trans. Ind. Appl.* **2013**, *49*, 1023–1030. [CrossRef]

9. Montoya, O.D.; Garcés, A.; Serra, F.M. DERs integration in microgrids using VSCs via proportional feedback linearization control: Supercapacitors and distributed generators. *J. Energy Storage* **2018**, *16*, 250–258. [CrossRef]

10. Gil-González, W.; Montoya, O.D.; Garces, A. Control of a SMES for mitigating subsynchronous oscillations in power systems: A PBC-PI approach. *J. Energy Storage* **2018**, *20*, 163–172. [CrossRef]

11. Shen, L.; Cheng, Q.; Cheng, Y.; Wei, L.; Wang, Y. Hierarchical control of DC micro-grid for photovoltaic EV charging station based on flywheel and battery energy storage system. *Electr. Power Syst. Res.* **2020**, *179*, 106079. [CrossRef]

12. Huang, C.N.; Chen, Y.S. Design of magnetic flywheel control for performance improvement of fuel cells used in vehicles. *Energy* **2017**, *118*, 840–852. [CrossRef]

13. Murad, M.A.A.; Milano, F. Modeling and Simulation of PI-Controllers Limiters for the Dynamic Analysis of VSC-Based Devices. *IEEE Trans. Power Syst.* **2019**, *34*, 3921–3930. [CrossRef]

14. Schiffer, J.; Seel, T.; Raisch, J.; Sezi, T. Voltage Stability and Reactive Power Sharing in Inverter-Based Microgrids With Consensus-Based Distributed Voltage Control. *IEEE Trans. Control Syst. Technol.* **2016**, *24*, 96–109. [CrossRef]

15. Joung, K.W.; Kim, T.; Park, J. Decoupled Frequency and Voltage Control for Stand-Alone Microgrid With High Renewable Penetration. *IEEE Trans. Ind. Appl.* **2019**, *55*, 122–133. [CrossRef]

16. Simpson-Porco, J.W.; Shafiee, Q.; Dörfler, F.; Vasquez, J.C.; Guerrero, J.M.; Bullo, F. Secondary Frequency and Voltage Control of Islanded Microgrids via Distributed Averaging. *IEEE Trans. Ind. Electron.* **2015**, *62*, 7025–7038. [CrossRef]

17. Garces, A. Uniqueness of the power flow solutions in low voltage direct current grids. *Electr. Power Syst. Res.* **2017**, *151*, 149–153. [CrossRef]

18. Montoya, O.D.; Gil-González, W.; Garces, A. Power flow approximation for DC networks with constant power loads via logarithmic transform of voltage magnitudes. *Electr. Power Syst. Res.* **2019**, *175*, 105887. [CrossRef]

19. Montoya, O.D.; Gil-González, W.; Garces, A. Optimal Power Flow on DC Microgrids: A Quadratic Convex Approximation. *IEEE Trans. Circuits Syst. II* **2019**, *66*, 1018–1022. [CrossRef]

20. Li, J.; Liu, F.; Wang, Z.; Low, S.H.; Mei, S. Optimal Power Flow in Stand-Alone DC Microgrids. *IEEE Trans. Power Syst.* **2018**, *33*, 5496–5506. [CrossRef]

21. Parhizi, S.; Lotfi, H.; Khodaei, A.; Bahramirad, S. State of the Art in Research on Microgrids: A Review. *IEEE Access* **2015**, *3*, 890–925. [CrossRef]

22. Montoya, O.D.; Gil-González, W.; Garces, A. Sequential quadratic programming models for solving the OPF problem in DC grids. *Electr. Power Syst. Res.* **2019**, *169*, 18–23. [CrossRef]

23. Montoya, O.D.; Gil-González, W.; Grisales-Noreña, L.F. Vortex Search Algorithm for Optimal Power Flow Analysis in DC Resistive Networks with CPLs. *IEEE Trans. Circuits Syst. II* **2019**. [CrossRef]

24. Garcés, A. On the Convergence of Newton's Method in Power Flow Studies for DC Microgrids. *IEEE Trans. Power Syst.* **2018**, *33*, 5770–5777. [CrossRef]

25. Montoya, O.D.; Garrido, V.M.; Gil-González, W.; Grisales-Noreña, L.F. Power Flow Analysis in DC Grids: Two Alternative Numerical Methods. *IEEE Trans. Circuits Syst. II* **2019**, *66*, 1865–1869. [CrossRef]

26. Yu, S.Y.; Kim, H.J.; Kim, J.H.; Han, B.M. SoC-Based Output Voltage Control for BESS with a Lithium-Ion Battery in a Stand-Alone DC Microgrid. *Energies* **2016**, *9*, 924. [CrossRef]

27. Robinson, S.; Papadopoulos, S.; Jadraque Gago, E.; Muneer, T. Feasibility Study of Integrating Renewable Energy Generation System in Sark Island to Reduce Energy Generation Cost and CO2 Emissions. *Energies* **2019**, *12*, 4722. [CrossRef]

28. Gil-González, W.; Montoya, O.D.; Garces, A. Modeling and control of a small hydro-power plant for a DC microgrid. *Electr. Power Syst. Res.* **2020**, *180*, 106104. [CrossRef]

29. Hassan, M.A.; Li, E.; Li, X.; Li, T.; Duan, C.; Chi, S. Adaptive Passivity-Based Control of dc–dc Buck Power Converter With Constant Power Load in DC Microgrid Systems. *IEEE J. Emerg. Sel. Top. Power Electron.* **2019**, *7*, 2029–2040. [CrossRef]

30. Shadmand, M.B.; Balog, R.S.; Abu-Rub, H. Model Predictive Control of PV Sources in a Smart DC Distribution System: Maximum Power Point Tracking and Droop Control. *IEEE Trans. Energy Convers.* **2014**, *29*, 913–921. [CrossRef]

31. Vafamand, N.; Khooban, M.H.; Dragičević, T.; Blaabjerg, F. Networked Fuzzy Predictive Control of Power Buffers for Dynamic Stabilization of DC Microgrids. *IEEE Trans. Ind. Electron.* **2019**, *66*, 1356–1362. [CrossRef]

32. Ghiasi, M.I.; Golkar, M.A.; Hajizadeh, A. Lyapunov Based-Distributed Fuzzy-Sliding Mode Control for Building Integrated-DC Microgrid With Plug-In Electric Vehicle. *IEEE Access* **2017**, *5*, 7746–7752. [CrossRef]

33. Mokhtar, M.; Marei, M.I.; El-Sattar, A.A. An Adaptive Droop Control Scheme for DC Microgrids Integrating Sliding Mode Voltage and Current Controlled Boost Converters. *IEEE Trans. Smart Grid* **2019**, *10*, 1685–1693. [CrossRef]

34. Liu, X.; Liu, Y.; Liu, J.; Xiang, Y.; Yuan, X. Optimal planning of AC-DC hybrid transmission and distributed energy resource system: Review and prospects. *CSEE J. Power Energy Syst.* **2019**, *5*, 409–422. [CrossRef]

35. Wogrin, S.; Gayme, D.F. Optimizing Storage Siting, Sizing, and Technology Portfolios in Transmission-Constrained Networks. *IEEE Trans. Power Syst.* **2015**, *30*, 3304–3313. [CrossRef]

36. Sachs, J.; Sawodny, O. A Two-Stage Model Predictive Control Strategy for Economic Diesel-PV-Battery Island Microgrid Operation in Rural Areas. *IEEE Trans. Sustain. Energy* **2016**, *7*, 903–913. [CrossRef]

37. Adefarati, T.; Bansal, R.C.; John Justo, J. Techno-economic analysis of a PV–wind–battery–diesel standalone power system in a remote area. *J. Eng.* **2017**, *2017*, 740–744. [CrossRef]

38. Moradi, M.; Abedini, M. A combination of genetic algorithm and particle swarm optimization for optimal DG location and sizing in distribution systems. *Int. J. Electr. Power Energy Syst.* **2012**, *34*, 66–74. [CrossRef]

39. Gandomkar, M.; Vakilian, M.; Ehsan, M. A Genetic–Based Tabu Search Algorithm for Optimal DG Allocation in Distribution Networks. *Electr. Power Components Syst.* **2005**, *33*, 1351–1362. [CrossRef]

40. Nekooei, K.; Farsangi, M.M.; Nezamabadi-Pour, H.; Lee, K.Y. An Improved Multi-Objective Harmony Search for Optimal Placement of DGs in Distribution Systems. *IEEE Trans. Smart Grid* **2013**, *4*, 557–567. [CrossRef]

41. Sultana, S.; Roy, P.K. Krill herd algorithm for optimal location of distributed generator in radial distribution system. *Appl. Soft Comput.* **2016**, *40*, 391–404. [CrossRef]

42. Grisales-Noreña, L.F.; Gonzalez Montoya, D.; Ramos-Paja, C.A. Optimal Sizing and Location of Distributed Generators Based on PBIL and PSO Techniques. *Energies* **2018**, *11*, 1018. [CrossRef]

43. Kaur, S.; Kumbhar, G.; Sharma, J. A MINLP technique for optimal placement of multiple DG units in distribution systems. *Int. J. Electr. Power Energy Syst.* **2014**, *63*, 609–617. [CrossRef]

44. Montoya, O.D.; Gil-González, W.; Grisales-Noreña, L. An exact MINLP model for optimal location and sizing of DGs in distribution networks: A general algebraic modeling system approach. *Ain Shams Eng. J.* **2019**. [CrossRef]

45. Montoya, O.D.; Gil-González, W.; Grisales-Noreña, L. Relaxed convex model for optimal location and sizing of DGs in DC grids using sequential quadratic programming and random hyperplane approaches. *Int. J. Electr. Power Energy Syst.* **2020**, *115*, 105442. [CrossRef]

46. Montoya, O.D. A convex OPF approximation for selecting the best candidate nodes for optimal location of power sources on DC resistive networks. *Eng. Sci. Technol. Int. J.* **2019**. [CrossRef]

47. Grisales-Noreña, L.F.; Garzon-Rivera, O.D.; Montoya, O.D.; Ramos-Paja, C.A. Hybrid Metaheuristic Optimization Methods for Optimal Location and Sizing DGs in DC Networks. In *Workshop on Engineering Applications*; Chapter Applied Computer Sciences in Engineering; Figueroa-García, J., Duarte-González, M., Jaramillo-Isaza, S., Orjuela-Cañon, A., Díaz-Gutierrez, Y., Eds.; Springer: Cham, Switzerland, 2019; Volume 1052, pp. 214–225. [CrossRef]

48. Montoya, O.D.; Garrido, V.M.; Grisales-Noreña, L.F.; Gil-González, W.; Garces, A.; Ramos-Paja, C.A. Optimal Location of DGs in DC Power Grids Using a MINLP Model Implemented in GAMS. In Proceedings of the 2018 IEEE 9th Power, Instrumentation and Measurement Meeting (EPIM), Salto, Uruguay, 14–16 November 2018; pp. 1–5. [CrossRef]

49. Maleki, A.; Pourfayaz, F.; Hafeznia, H.; Rosen, M.A. A novel framework for optimal photovoltaic size and location in remote areas using a hybrid method: A case study of eastern Iran. *Energy Convers. Manag.* **2017**, *153*, 129–143. [CrossRef]

50. Montoya, O.D.; Gil-González, W.; Grisales-Noreña, L.; Orozco-Henao, C.; Serra, F. Economic Dispatch of BESS and Renewable Generators in DC Microgrids Using Voltage-Dependent Load Models. *Energies* **2019**, *12*, 4494. [CrossRef]

51. Simpson-Porco, J.W.; Dörfler, F.; Bullo, F. On Resistive Networks of Constant-Power Devices. *IEEE Trans. Circuits Syst. II* **2015**, *62*, 811–815. [CrossRef]

52. Zhang, L. Artificial Neural Network Architecture Design for EEG Time Series Simulation Using Chaotic System. In Proceedings of the 2018 Joint 7th International Conference on Informatics, Electronics Vision (ICIEV) and 2018 2nd International Conference on Imaging, Vision Pattern Recognition (icIVPR), Kitakyushu, Japan, 25–29 June 2018; pp. 388–393. [CrossRef]

53. Du, Y.C.; Stephanus, A. Levenberg-Marquardt Neural Network Algorithm for Degree of Arteriovenous Fistula Stenosis Classification Using a Dual Optical Photoplethysmography Sensor. *Sensors* **2018**, *18*, 2322. [CrossRef]

54. Wilson, P.; Mantooth, H.A. Chapter 10—Model-Based Optimization Techniques. In *Model-Based Engineering for Complex Electronic Systems*; Wilson, P., Mantooth, H.A., Eds.; Newnes: Oxford, UK, 2013; pp. 347–367. [CrossRef]

55. Naghiloo, A.; Abbaspour, M.; Mohammadi-Ivatloo, B.; Bakhtari, K. GAMS based approach for optimal design and sizing of a pressure retarded osmosis power plant in Bahmanshir river of Iran. *Renew. Sustain. Energy Rev.* **2015**, *52*, 1559–1565. [CrossRef]

56. Montoya, O.D. Solving a Classical Optimization Problem Using GAMS Optimizer Package: Economic Dispatch Problem Implementation. *Ingenieria y Ciencia* **2017**, *13*, 39–63. [CrossRef]

57. GAMS Development Corp. GAMS Free Demo Version. Available online: https://www.gams.com/download/ (accessed on 14 February 2020).

Article

A Simplified Method to Avoid Shadows at Parabolic-Trough Solar Collectors Facilities

Nuria Novas [1], Aránzazu Fernández-García [2] and Francisco Manzano-Agugliaro [1,*]

[1]   Department of Engineering, University of Almeria, ceiA3, 04120 Almeria, Spain; nnovas@ual.es
[2]   CIEMAT-Plataforma Solar de Almería, Ctra. Senés, 04200 Tabernas, Spain; afernandez@psa.es
*   Correspondence: fmanzano@ual.es; Tel.: +34-950-015396; Fax: +34-950-015491

Received: 10 January 2020; Accepted: 6 February 2020; Published: 13 February 2020

Abstract: Renewable energy today is no longer just an affordable alternative, but a requirement for mitigating global environmental problems such as climate change. Among renewable energies, the use of solar energy is one of the most widespread. Concentrating Solar Power (CSP) systems, however, is not yet fully widespread despite having demonstrated great efficiency, mainly thanks to parabolic-trough collector (PTC) technology, both on a large scale and on a small scale for heating water in industry. One of the main drawbacks to this energy solution is the large size of the facilities. For this purpose, several models have been developed to avoid shadowing between the PTC lines as much as possible. In this study, the classic shadowing models between the PTC rows are reviewed. One of the major challenges is that they are studied geometrically as a fixed installation, while they are moving facilities, as they have a tracking movement of the sun. In this work, a new model is proposed to avoid shadowing by taking into account the movement of the facilities depending on their latitude. Secondly, the model is tested to an existing facility as a real case study located in southern Spain. The model is applied to the main existing installations in the northern hemisphere, thus showing the usefulness of the model for any PTC installation in the world. The shadow projected by a standard, the PTC (S) has been obtained by means of a polynomial approximation as a function of the latitude (Lat) given by $S = 0.001 - \text{Lat}^2 + 0.0121 - \text{Lat} + 10.9$ with R^2 of 99.8%. Finally, the model has been simplified to obtain in the standard case the shadows in the running time of a PTC facility.

Keywords: CSP; PTC rows; solar; shadowing; energy; renewable energy

---

1. Introduction

The continued increase in energy demands worldwide is leading in an emergently unsustainable situation [1], then energy-related greenhouse gas emissions will result in substantial climate change if no decisive action is taken to reduce global warming. With the United Nations, this target to hold means global temperature will rise by the end of the century to at least 2 °C [2]. These facts represent an essential driving force for the gradual implementation of safe and feasible alternatives in all power-consuming sectors [3] in addition to policies that help industries implement strategies to improve the efficiency of energy use through innovative technologies. Where they are integrated within national and foreign policies, and with the mechanisms of ecological technological innovation to give demand to energy saving and emissions reduction [4]. Today, not only is it being applied to the industrial sector, but citizen awareness has brought these energy technology innovations into the home. There are examples of the active use of smart technologies with the internet of things for the home, as in [5] with the aim of managing energy performance and optimizing consumption and obtaining net zero energy. The internet and smart phones have enabled real-time monitoring of sensors and actuators that control active consumption in homes.

The society and energy system as a whole need to be more energy efficient, and the development of renewable energy sources can help. One of the problems posed by renewable sources is the balance between production and consumption, for which an accurate prediction of the load and a correctly sized storage system are desirable. In [6], the authors have studied two situations for distributed photovoltaic production: "those that predict by separating the load portion due to consumption habits from the production portion due to local climatic conditions, and those that try to predict the load as a whole". Predicting the behaviour of the grid and its dependence on climatic conditions largely defines the efficiency of the system. Precise forecasting techniques are tools for integrating renewable energy systems into the power grid [7]. However, it attention should be given to their rate of development, rapidly growing share in energy demand, and impact in the market [8]. Solar energy is a renewable source that is clean, inexhaustible, and allows for local energy independence [9]. The total energy production from the sun is 3.8×1020 MW, equivalent to 63 MW/m2 of solar surface; only a small fraction, 1.7×1014 kW, of the total emitted radiation is captured by the Earth [10]. Although it is estimated that even with this tiny portion, 30 min of solar radiation reaching the Earth is equal to the worldwide energy requirement for a year [11]. Scientific research has provided a high importance to solar energy in the last three decades, being the second most important renewable resource, with the 26% of the scientific publications [12].

The relatively low flow of solar energy received at the Earth's surface can be surpassed by the use of concentrating solar collectors that transform solar energy into other types of energy, usually thermal in Concentrating Solar Power (CSP) [13] and Central Receiver System (CRS) [14]. The concentration of solar radiation by reflective mirrors on the receptor of a Thermal Conversion System has the important advantage of reducing thermal energy losses compared to unconcentrated systems [15], resulting in increased thermal conversion performance for the operating conditions specified and allowing higher working temperatures to be achieved with appropriate efficiencies [16]. The thermal energy collected on the receiver of the concentrating solar systems is typically used to produce electricity through a conventional power block, in the commonly named concentrating solar power plants. Although the basic technology had been under development for about 140 years, solar thermal electricity (STE) on grid was not achieved until the 1980s [17].

The parabolic-trough solar energy technology is the most tested and the cheapest large-scale solar energy technology available today for the use of thermal solar energy with different types of working fluids [18,19]. The improvements in these parabolic-trough systems are still far from complete. The introduction of internal longitudinal fins and a reflector shield together results in a thermal efficiency improvement of 2.41% compared to the same system without the improvements [20].

The electrical capacity of CSP plants currently in operation worldwide is 5.7 GW by the end of 2020, according to the International Energy Agency's (IEA) forecast for 2050, an 11% of the worldwide energy mix will be provided by CST systems [16]. 84% of this power is produced by plants with parabolic-trough collectors (PTCs) [21]. Although these solar concentration technologies are among the most widely implemented worldwide, alternative systems are still being studied [22,23].

PTCs are integrated by a trough-shaped reflector with a parabolic cross-section that concentrates and focuses the direct solar irradiation in parallel to the axis of the collector in a focal line (see Figure 1). A receiving pipe with a fluid that flows inside it and which absorbs the concentrated solar energy from the pipe walls and increases its enthalpy is placed along the length of the collector at its focus. The tube is typically coated with a selective layer to reduce thermal losses by radiation to the ambient. A cylindrical glass enclosure concentric to the receiver pipe is also employed to reduce thermal heat loss by convection into the environment. A single-axis tracking system turns the collector to be sure that the sun's ray drops parallel to the collector's axis.

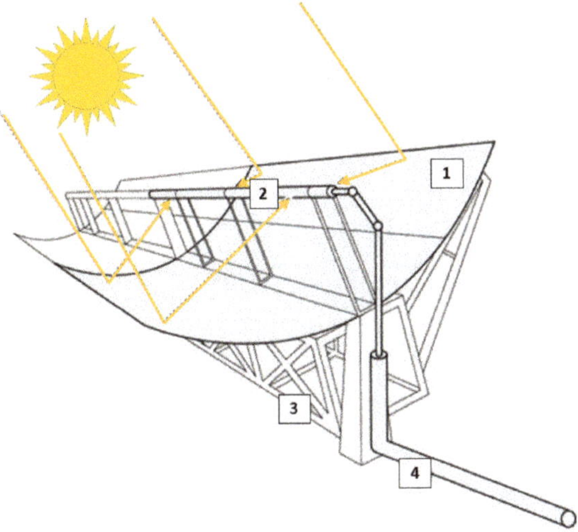

Figure 1. Scheme of working of a PTC facility. (1) Reflector; (2) Absorber tube; (3) Structure; (4) Solar Field Piping.

The solar installation is intended to be modular in design and consists of several parallel rows of solar collectors [18]. It involves a large number of reflecting surfaces, from 0.6 to 10 ha/MWe, depending on the capacity of the storage and auxiliary systems [24]. The different rows are separated among them to avoid shadowing and permit the access and handling of cleaning devices. Collector shadowing means a reduction in the net aperture area, so reducing the amount of thermal energy that can be supplied by the solar field. In this sense, distance between adjacent collector rows should be as high as possible.

In general, the land occupation factor (that is, the aperture area of the solar field divided by the land surface occupied by the whole plant) is around 0.245 [25]. This means that the land area required to install a plant is around four times the solar field area, partially due to the separation among solar collector rows. Hence, this separation should be minimized to avoid an unreasonable land use. Consequently, an optimization process to calculate the collector-row separation is required, searching for a compromise that maximizes separation to reduce shadowing but minimizes it to use land wisely. The effectiveness of this optimization process depends on the method used to calculate the shadowing between adjacent PTCs.

2. Classical Methods for the Sizing of PTC: A Brief Overview

The designs of solar energy installations should be designed for the most efficient use of energy. In a classical model, the area of the solar collector should be perpendicular to the received sunlight. However, given the Earth's declination, the relative positions of the Earth's hemispheres vary continuously in relation to the sun throughout the year and therefore the day. Therefore, in order to get the solar rays perpendicular to each PTC, the tilt of a solar installation with respect to the horizon should also change throughout the year. Thus, a common solution to maximize energy generation is getting the solar installation in the most perpendicular position to the sun at the time of the winter solstice.

It is known that the time of the zenithal passing by of the sun or meridian of the location, i.e., the actual 12 h of the solar day, establishes the relationship between latitude (Φ), height of the sun on the horizon (h) and declination angle (δ). See Figure 2, which is provided by the following equation [26]:

$$\gamma s = (\pi/2) - |\delta| - \Phi \tag{1}$$

$$\theta_{ZS} = (\pi/2) - \gamma s = |\delta| + \Phi \tag{2}$$

where δ is the Earth's decline (at the winter solstice), γs is solar altitude angle and θ_{ZS} as zenith angle.

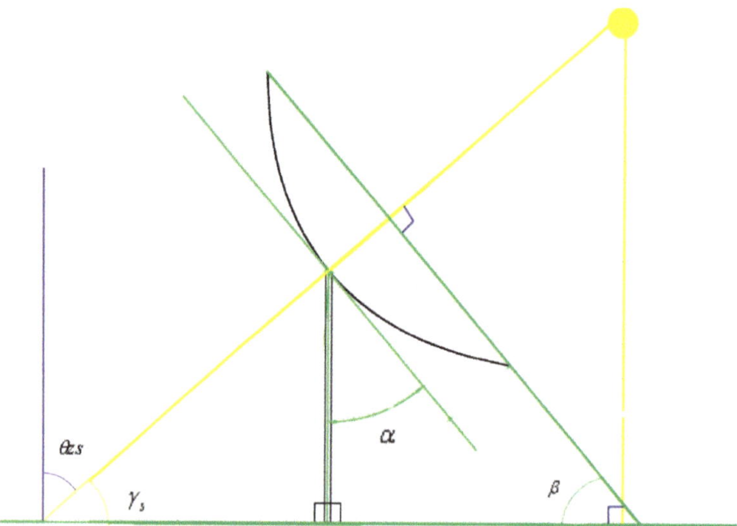

Figure 2. Geometry on a PTC for the sun's rays (section view).

3. Standard Methods for Determining the Spacing between Collectors in PTC Facilities

3.1. Standard Method 1

It is based on the calculation of the distance (D) between the PTCs as a function of the height of the sun (h) for whom the facility was designed. Figure 3 shows the geometry to derive Equation (4):

$$D = d + d' = W \cdot \frac{cos\alpha}{tg\,h} + W\,sin\,\alpha \tag{3}$$

$$D = W \cdot \left(\frac{cos\alpha}{tg\,h} + sin\,\alpha \right) \tag{4}$$

where W is the width of opening plane and α is the tilt angle relative to the vertical of the collector (azimuth of the panel). α is calculated to achieve sun rays perpendicular- Collector along the entire operating time of the solar plant. α is the solar tracking parameter which varies continuously all the time depending on the time, day, and location of the PTC facility.

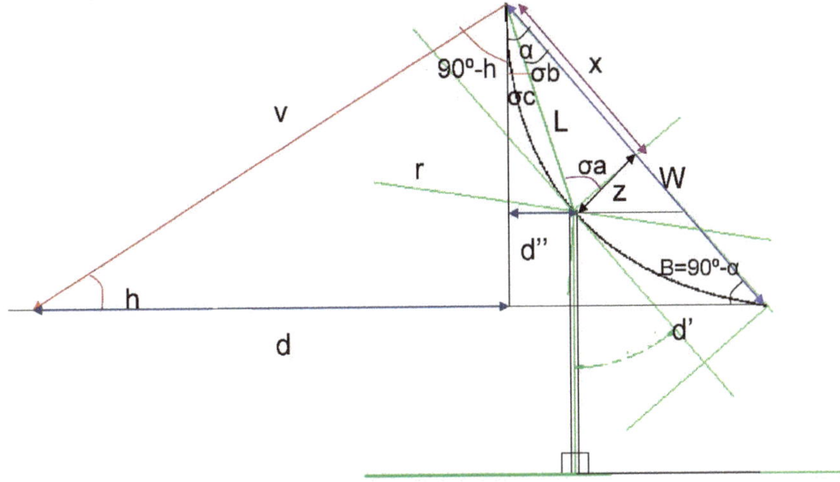

Figure 3. Geometry on a PTC for the sun's rays (frontal view).

In this way, the shadow gets a horizontal spacing D since the first line of the PTC being h > h′, see Figure 4, and so the second line must at least be placed in the PTC2h. In case the sun gets a height of h′, the shading would have a horizontal spacing D′, and the second PTC line would be positioned in PTC2h′.

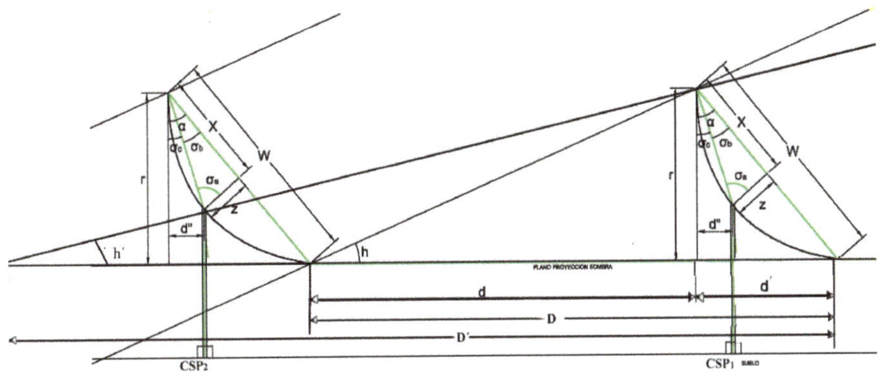

Figure 4. Row alignment minimum PTC depending on the solar altitude (Standard Method 1).

Figure 4 shows the shadow geometry of the first PTC line (PTC 1) over the second PTC line (PTC 2 h).

3.2. Standard Method 2

This other standard method is used mainly in Spain, so that four hours of sunshine are ensured around midday on the winter solstice. In this way, in place of estimating the positions for a particular sun height, the latitude of the place is the necessary data for the use of this other standard method. It is obtained that the measured distance across the rows (d) of the PTCs of height W′ is shown at Equation (5) (Figure 5) [27].

$$d \leq W' \cdot k \tag{5}$$

where k is the zero-dimensional factor, which varies according to the latitude of the place

$$k = \frac{1}{tg\,(61^{\circ} - \Phi)} \tag{6}$$

and Φ as latitude in (°).

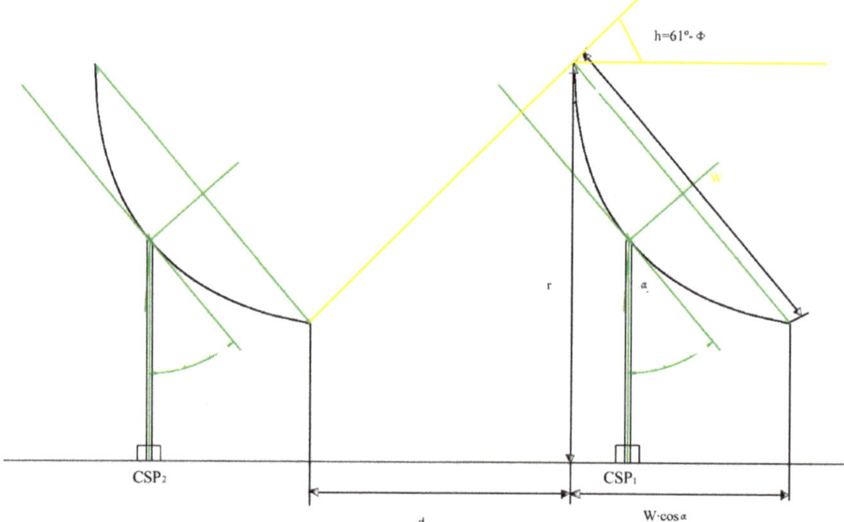

Figure 5. Minimum PTC row alignment as solar altitude function (standard method 2).

The spacing achieved (d) should be added to the horizontal length collector spacing at an inclination angle (α), W·sin α, as reported on Equation (7) [28].

$$D = d + d' = \frac{W'}{tg\,(61^{\circ} - \Phi)} + W\sin\,\alpha = W\left(\frac{\cos\,\alpha}{tg\,(61^{\circ} - \Phi)} + \sin\,\alpha\right) \tag{7}$$

4. Proposed Method

The proposed method provides for the estimation of the accurate shadowing projection of PTCs for every solar hour. This process, it should be used to estimate the optimal use of area in accordance with the energy needs of PTC installations at a given location for a latitude done (Φ) based on the amount of solar gain and the inclination of the collector (α, angle relative to the horizontal PTC panel β).

In this way, the shadow projected on the ground for each PTC can be calculated using three directions: north, east, and west. For this, it is necessary to know the azimuth of the sun at the winter and summer solstices.

With the geometry and the tilt angle of the collector and using the data from the geometrical relations, shadows can be calculated for both of the corners of each PTC, and thus the surrounding polygon of the shadow path as a maximum area. Therefore, the envelope will be the exterior silhouette that forms the shadow for the period studied (see Figure 6). Thus, it is possible to calculate the minimum distance between the rows of PTCs avoid the effect of shadows in the period studied.

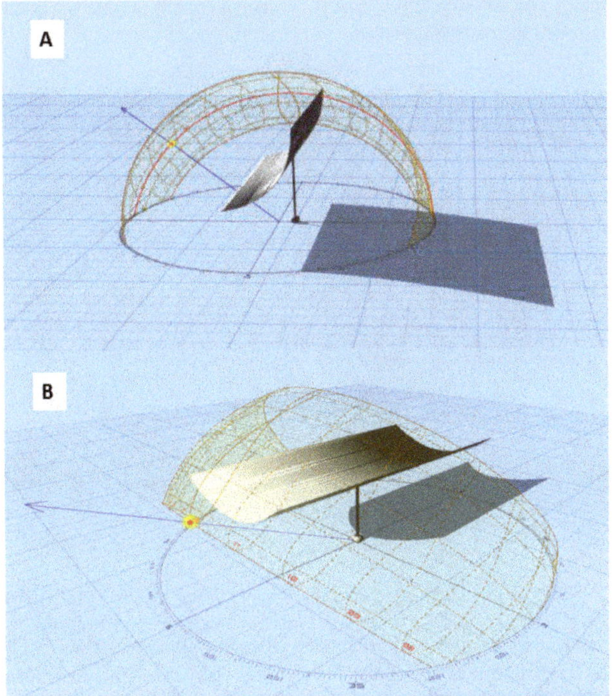

Figure 6. Proposed method: isometric view of PTC. (**A**) Shadowing at 10 a.m.; (**B**) Shadowing at 12 a.m.

Figure 7 shows a flowchart of the methodology followed where from the data of the PTC field location and its dimensions and inclination, all the necessary data for the calculation of the distance between pylons without shadowing are calculated.

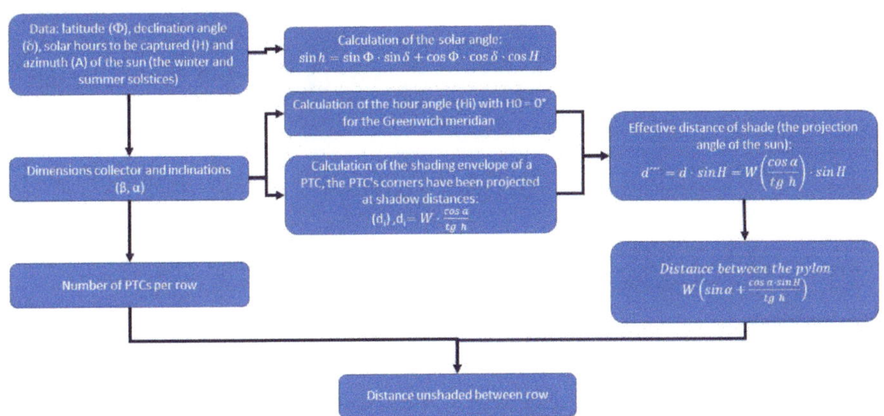

Figure 7. Flowchart of the methodology for the calculation of the distance of a PTC row.

4.1. Solar Angle Calculation

At a specific latitude, the height of the sun depends on the hour. To establish the height of the sun-latitude relationship, basic knowledge of celestial physics is used, where the planet Earth is located in the centre (Figure 8). The equatorial plane of the celestial sphere (NS) is the equatorial plane of the Earth, where the azimuth is positive when viewed from the north (A), A* = 360° − A, i.e., clockwise.

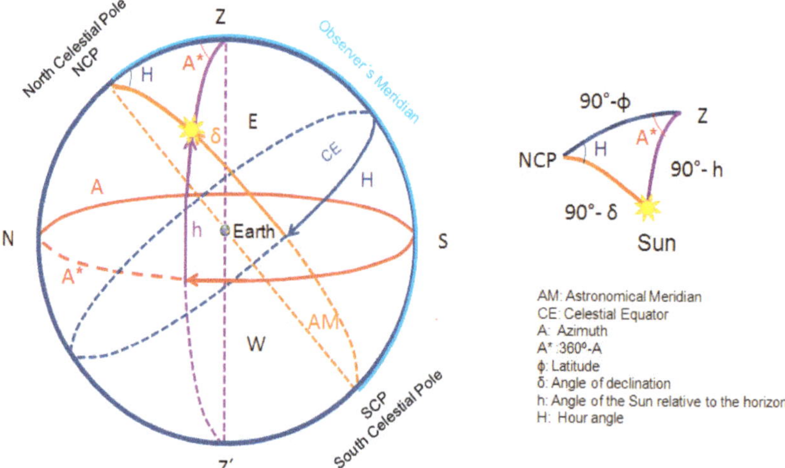

Figure 8. Spherical triangle model for the estimation of the sun's shadow at a specific latitude.

The angle of the astronomical meridian with the equatorial plane is the solar height (h), Φ is the latitude, and δ is the declination angle of the Earth. Therefore, to determine the solar height (h) at a particular time (H) at some location on Earth (latitude), for a spherical triangle, the corresponding equations of spherical trigonometry are employed logically, as presented in Figure 7.

If a spherical triangle is used, the Z point will be the origin of the coordinates. Note that the sides are the angles in radians. Therefore, in our case, the sides of the triangle are:

$(90° − δ) \Rightarrow$ hour angle (H)

$(90° − Φ) \Rightarrow$ the azimuth supplement (A*)

$(90° − h) \Rightarrow$ solar height (h)

Using the law of sines to link all these variables in one system of equations, the first two sides and their angles can be replaced according to Equation (8) [28]:

$$\frac{\left(90° − δ\right)}{\sin(A*)} = \frac{\left(90° − h\right)}{\sin H} \tag{8}$$

$$\frac{\sin\left(90° − δ\right)}{\sin\left(360° − A\right)} = \frac{\sin\left(90° − h\right)}{\sin H} \Longrightarrow \frac{\cos δ}{−\sin A} = \frac{\cos h}{\sin H} \tag{9}$$

In the last two fractions, the sine of the 90° and 360° angles are 1 and 0, respectively. In addition, sin (90° − v) = cos v, sin (360° − r) = −sin (r), giving the next equation:

$$\cos δ · \sin H = −\sin A · \cos h \tag{10}$$

Using spherical trigonometry, i.e., the first law of cosines next to it (90° − δ), the consequent equation is:

$$\sin δ = \sin Φ · \sin h + \cos Φ · \cos h · \cos A \tag{11}$$

Then, in spherical trigonometry, if the first law of cosines is applied for (90° − h), the next equation can be found:

$$\sin h = \sin \Phi \cdot \sin \delta + \cos \Phi \cdot \cos \delta \cdot \cos H \tag{12}$$

Thus, the value of h is obtained in terms of the variables (δ, Φ, H).

On the other hand, it is also necessary to know the measures and inclinations (β, α) of the PTC to calculate the shadow. To establish the shading end points, we must consider at first that the solar hours are used for the design of the PTC facility. The solar hours refer to the central peak hour (H_0) and the hours that are equally divided backwards and forwards from this central hour. The central peak is assumed to be $H_0 = 0°$ for the Greenwich meridian, with each hour corresponds to 15 degrees, with the adding on the right of 15 degrees per hour of solar gain and the subtracting on the left of 15° per hour (with 0 assumed to be 360° to prevent values of angles as negative). e.g., in Spain, for a setting of four hours of sun, at the time of 10:00 h, there would be an H of 330°, and at 12:00 h, it is reached an H of 30° (see Figure 9 as guidance).

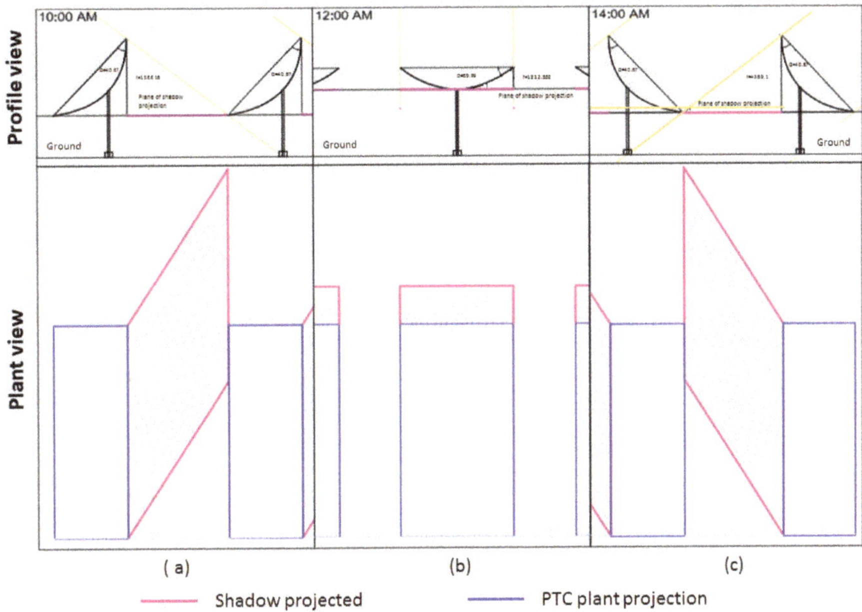

Figure 9. Shadows. (**a**) Shadow at 10:00, (**b**) shadow at 12:00; and (**c**) Shadow at 14:00.

4.2. The Extent of the Shade

For the estimation of the shadow, the four corners of the concentrator are chosen. So, with known values of δ, Φ, H, and using Equation (12), it can be calculated the projected shadow (see Figure 10), where the shadow is calculated at 10, 12, and 14 h in continuous time. The outer contour of both shadows will be the envelope, these are shown in Figure 10, where the contours of full shadow at 10, 12, and 14 h have been represented.

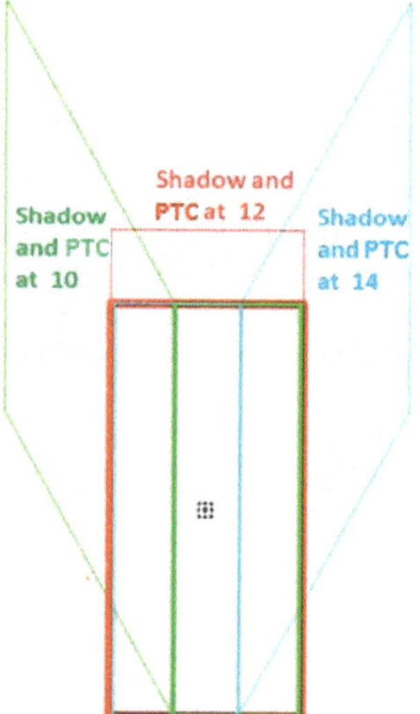

Figure 10. Range of the total projection of the collector's shadow.

Afterward, the distance (d) to the shading for every point is calculated based on Equation (12), linking the tilt angle of the PTC corner, the length, and the sun height.

$$d = W \cdot \frac{\cos \alpha}{\operatorname{tg} h} \tag{13}$$

Until now, the PTC's corners have been projected at shadow distances (d_i). To plot these shading distances on the floor, polar coordinates are applied, whereby the angle H_i is the hourly angle of the point i. The critical event is the shortest day of one year, where h represents the shortest day of the solar field design. The shading envelope of a PTC is obtained, and from this envelope, the shadow projected for each whole row of the PTC assembly can be drawn so that no shadow can be cast between the rows.

The data to be computed is the spacing between the rows of PTC lines or the distance between the pylon (see Figure 11), the triangle must be solved, whose vertexes are: shadow of the first hour, shadow of the last hour (third point), and the projection of the corner of PTC. The vertex angle P is (360 − (H_1 − H_3)), meaning the difference in the hour angles of the other two vertexes, and the sides from point P to first point and to third point are d_1 and d_3, respectively. Before for h_1 and h_3 are calculated the values of d_1 and d_3. Then, it is possible to determine an effective distance of shade considering the projection angle of the sun, as shown in Equation (14) (Figure 11),

$$d''' = d \cdot \sin H = W \left(\frac{\cos \alpha}{\operatorname{tg} h} \right) \cdot \sin H \tag{14}$$

where the distance between the pylon can be obtained depending on the dimension's collector, the angle from the vertical of the collector, the solar hour angle, and height, as shown in the following Equation (16). In short, the distance depends the dimensions of the collector and the location of the solar field.

$$\text{Distance between the pylon} = d''' + d' = W \left(\frac{\cos \alpha \cdot \sin H}{\operatorname{tg} h} \right) + W \sin \alpha \tag{15}$$

$$\text{Distance between the pylon} = W \left(\sin \alpha + \frac{\cos \alpha \cdot \sin H}{\operatorname{tg} h} \right) \tag{16}$$

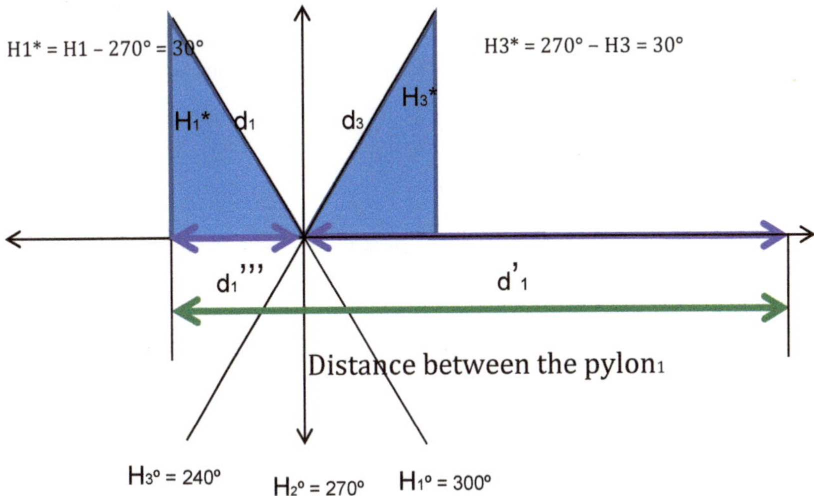

Figure 11. Diagram of d''' calculation polygons.

The hour angle (H) corresponds to the position of the observer with respect to the sun and the azimuth angle (A) is based on the position of the observer with respect to the north. Then, by using Equation (12), shadows at sunrise are calculated, Ortho, h = 0°, that is, sin h = 0; where for the equations, the angles are used in radians, whereas the solutions are expressed as degrees (°) to make it more user friendly (Equation (17)).

$$0 = tg\ \Phi \cdot tg\ \delta + \cos H \tag{17}$$

The result of the calculations in $H_{ORTHO}°$ and $H_{SUNSET}°$.

Considering the opening plane size as standard one, $W = 5.760$ m and focal distance $f = 1.710$ m for the distance calculation. Known W and f, it is possible the calculation of the distance of the vertex of the collector perpendicular to the aperture plane according to Equation (18):

$$z = \frac{\left(\frac{W}{2}\right)^2}{4f} \tag{18}$$

5. Results and Discussion

5.1. Case Study: Results

The case study will be the most unfavourable day, i.e., on December 22, the winter solstice. The shadow projected by each PTC was estimated, allowing to determine the minimum distance of the next row of PTC. The first case of study was situated in the southern of Spain, CIEMAT-PSA research centre (latitude 37.091° N; longitude 2.355° W). The data used were declination $\delta = -23°\ 27'$; latitude $\Phi = 37.093°$ N, for Equations (11) and (12). It is estimated that in this location the PTCs do not reach adequate temperature to start working until two hours after sunrise. Therefore, in this case, the shadows will be calculated for the period of time in which the installation is in service. That is, from 10 to 14 h.

H_{ORTHO} and H_{SUNSET} are calculated using Equation (17), obtaining $H_{ORTHO} = 289.09°$ and $H_{SUNSET} = 70.91°$.

Table 1 shows the results obtained for each shadow of the three points (first and third) as shown in Figures 9 and 10. These outputs are considered to be valid for every point in time at which the shading distance was calculated. The known angles $H_1 = 330°$; $H_3 = 30°$, angles (h_1, h_3), and distances d (d_1 y d$_3$) are computed, considering the dimensions of a standard PTC of the commonly used Eurotrough model [29] (the aperture plane size W = 5.760 m and focal distance f = 1.710 m) for the calculation of the distance of the vertex of the collector perpendicular to the aperture plane according to Equation (18), resulting in that z = 1.212 m.

Table 1. Calculations for every shadow of the three points at south of Spain on 22 December 2019 (latitude 37.091° N).

Solar Hour	Time Angle (° (H)	Elevation or Solar Height (°) (h)	Flat Tilt Opening (°) (α)	Distance between Pylons (m) d''' + d' = W (sinα + cosα (sin H/tg h))
10:00:00	330.000	0.416	40.666	8.847
12:00:00	300.000	0.500	89.999	5.760
14:00:00	270.000	0.583	40.665	8.847

As can be observed in the results presented in Table 1, the distance calculated between PTC pylons shows a perfect symmetry throughout the day with respect to the moon, as was expected. This fact proves that a first and essential requirement to check the validity of the proposed model is accomplished.

5.2. Extension of the Case Study to Worldwide

The proposed model has been calculated in several key locations for PTC facilities in the northern hemisphere, from a latitude of 14 degrees to almost 51 degrees. Table 2 summarizes the results obtained. Clearly, the distance increases with increasing latitude.

Table 2. Calculations according to the proposed modelling the main PTC facilities in the northern hemisphere.

Country	Emplacement	Latitude (° N)	Solar Hour (2 h after Sunrise)	Calculated Shadow Distance d''' + d' = W (sinα + cosα (sin H/tg h)) (m)
Thailand	Kanchanaburi	14.022	8:25:07	11.294
USA	Kailua-Kona (Hawai)	19.639	8:35:43	11.546
UEA	Medinat Zayed (Abu Dabi)	23.660	8:43:00	11.749
USA	Indiantown (Florida)	27.027	8:51:07	11.957
Algeria	HassiR'mel	32.928	9:05:07	12.407
Morocco	Ain Beni Mathar (Oujda)	34.088	9:08:00	12.508
USA	Mojave Desert (California)	35.031	9:10:00	12.593
Spain	Almeria	37.051	9:16:25	12.787
Italy	Massa Martana	42.776	10:08:59	13.374
Canada	Kingsey Falls (Québec)	45.860	9:46:00	13.699
Germany	Jülich	50.922	10:08:59	14.147

If, from the data obtained in Table 2, a simple model is established to calculate the shadow of a standard concentrator (the opening plane size W = 5.760 m and focal distance f = 1.710 m), where S is the calculated shadow and Lat is the northern latitude, the following models are obtained:

The Linear Estimation:

$$S = 0.0796 \text{ Lat} + 9.9243 \tag{19}$$

with $R^2 = 0.9769$

The Polynomial Estimation:

$$S = 0.001 \text{ Lat}^2 + 0.0121 \text{ Lat} + 10.9 \tag{20}$$

with $R^2 = 0.9984$

If the solutions that would be obtained with each model are represented. Figure 12 is obtained, where, the area marked in green, which is to say between 20 and 45 degrees north latitude, there is scarce difference between both models. Outside this area, the linear model underestimates the magnitude of the shadow and therefore its use would not be advisable. For example, at 14 degrees north latitude, the linear model underestimates the shadow by 25 cm, while the polynomial model underestimates it by less than 3 cm. In short, the results suggest the use of the polynomial model obtained for the calculation of the shadows since it offers very good results as it has an R^2 greater than 99.8%.

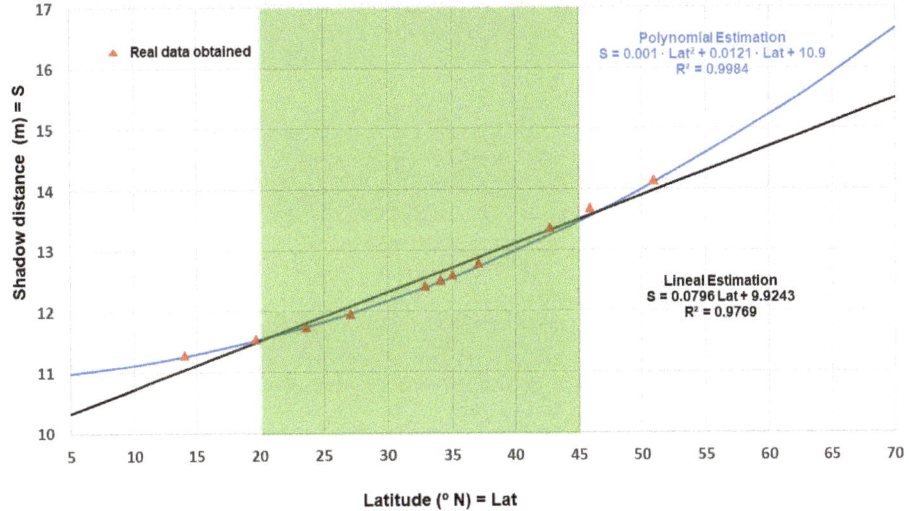

Figure 12. Different models obtained for the calculation of shadows in depending on the latitude.

If the maximum value obtained for the distance between the PTC pylons is noted, a separation of around 11 to 14 m must be selected for the layout of the solar field during the design phase. This would involve a significantly lower land occupation (around 50% lower) compared to the total area to be covered if the thumb rule (of four times the aperture area) is considered. Consequently, the model presented is this work is a very useful tool for CSP plant designer because it is easy to apply, and the investment costs are significantly reduced thanks to the reduction in the land occupation for the solar field.

6. Conclusions

In this work, a new methodology has been proposed for calculating the shadows of parabolic-trough solar collectors, PTC, depending on the geographical latitude of the CSP plant location. The latitude and the standard dimensions of a standard PTC has been considered. In addition, the distance between PTC suggested was calculated by estimating that the start-up of the plant is done around two hours after the sunrise. Since the model developed, although not complex but needs quite a lot of calculations, an approximate model has been calculated for this type of standard CSP, obtaining a linear with R^2 of 97.69% and a polynomial model with R^2 of 99.8%. Both run well within the range of 20 to 45 degrees latitude, but outside this zone, the polynomial model works best. In short, it is proposed to use the polynomial model obtained. Furthermore, this work opens new perspectives for the calculation of shadows in CSPs plants since the methodology developed in this work can be used to establish simple shadow calculation models when the dimensions of the PTC are different from the one used in this work (or even if other type of CSP collectors are studied) or when the operating times of the installation are different.

Author Contributions: N.N., A.F.-G. and F.M.-A. conceived, designed the research and wrote the manuscript. All authors have read and agreed to the published version of the manuscript.

Funding: No external funding was received for this research.

Acknowledgments: The authors would like to thank to the CIAIMBITAL (University of Almeria, CeiA3) for its support. The manuscript was funded by I+D+I Project UAL18-TIC-A025-A, University of Almeria, the Ministry of Economy, Knowledge, Business and University and the European Regional Development Fund (FEDER).

Conflicts of Interest: The authors declare no conflict of interest. The funders had no role in the design of the study; in the collection, analyses, or interpretation of data; in the writing of the manuscript; or in the decision to publish the results.

References

1. Baños, R.; Manzano-Agugliaro, F.; Montoya, F.; Gil, C.; Alcayde, A.; Gómez, J. Optimization methods applied to renewable and sustainable energy: A review. *Renew. Sustain. Energy Rev.* **2011**, *15*, 1753–1766. [CrossRef]
2. Schaeffer, M.; Hare, W.; Rahmstorf, S.; Vermeer, M. Long-term sea-level rise implied by 1.5 °C and 2 °C warming levels. *Nat. Clim. Chang.* **2012**, *2*, 867–870. [CrossRef]
3. Meyers, S.; Schmitt, B.; Vajen, K. The future of low carbon industrial process heat: A comparison between solar thermal and heat pumps. *Sol. Energy* **2018**, *173*, 893–904. [CrossRef]
4. Miao, C.; Fang, D.; Sun, L.; Luo, Q.; Yu, Q. Driving effect of technology innovation on energy utilization efficiency in strategic emerging industries. *J. Clean. Prod.* **2018**, *170*, 1177–1184. [CrossRef]
5. Alfaris, F.; Juaidi, A.; Manzano-Agugliaro, F. Intelligent homes' technologies to optimize the energy performance for the net zero energy home. *Energy Build.* **2017**, *153*, 262–274. [CrossRef]
6. Massidda, L.; Marrocu, M. Decoupling Weather Influence from User Habits for an Optimal Electric Load Forecast System. *Energies* **2017**, *10*, 2171. [CrossRef]
7. Massidda, L.; Marrocu, M. Use of Multilinear Adaptive Regression Splines and numerical weather prediction to forecast the power output of a PV plant in Borkum, Germany. *Sol. Energy* **2017**, *146*, 141–149. [CrossRef]
8. Islam, T.; Huda, N.; Abdullah, A.; Saidur, R. A comprehensive review of state-of-the-art concentrating solar power (CSP) technologies: Current status and research trends. *Renew. Sustain. Energy Rev.* **2018**, *91*, 987–1018. [CrossRef]
9. Hernandez-Escobedo, Q.; Rodriguez-Garcia, E.; Saldaña-Flores, R.; Fernández-García, A.; Manzano-Agugliaro, F. Solar energy resource assessment in Mexican states along the Gulf of Mexico. *Renew. Sustain. Energy Rev.* **2015**, *43*, 216–238. [CrossRef]
10. Sansaniwal, S.K.; Sharma, V.; Mathur, J. Energy and exergy analyses of various typical solar energy applications: A comprehensive review. *Renew. Sustain. Energy Rev.* **2018**, *82*, 1576–1601. [CrossRef]
11. Sengupta, M.; Xie, Y.; Lopez, A.; Habte, A.; Maclaurin, G.; Shelby, J. The National Solar Radiation Data Base (NSRDB). *Renew. Sustain. Energy Rev.* **2018**, *89*, 51–60. [CrossRef]
12. Manzano-Agugliaro, F.; Alcayde, A.; Montoya, F.; Zapata-Sierra, A.J.; Gil, C. Scientific production of renewable energies worldwide: An overview. *Renew. Sustain. Energy Rev.* **2013**, *18*, 134–143. [CrossRef]
13. Cruz-Peragón, F.; Palomar, J.; Casanova, P.; Dorado, M.; Manzano-Agugliaro, F. Characterization of solar flat plate collectors. *Renew. Sustain. Energy Rev.* **2012**, *16*, 1709–1720. [CrossRef]
14. Behar, O.; Khellaf, A.; Mohammedi, K. A review of studies on central receiver solar thermal power plants. *Renew. Sustain. Energy Rev.* **2013**, *23*, 12–39. [CrossRef]

15. Fernández-García, A.; Rojas, E.; Pérez, M.; Silva, R.; Hernandez-Escobedo, Q.; Manzano-Agugliaro, F.; Pérez-García, M. A parabolic-trough collector for cleaner industrial process heat. *J. Clean. Prod.* **2015**, *89*, 272–285. [CrossRef]

16. Sutter, F.; Fernández-García, A.; Wette, J.; Reche-Navarro, T.J.; Martínez-Arcos, L. Acceptance criteria for accelerated aging testing of silvered-glass mirrors for concentrated solar power technologies. *Sol. Energy Mater. Sol. Cells* **2019**, *193*, 361–371. [CrossRef]

17. Kasaeian, A.; Nouri, G.; Ranjbaran, P.; Wen, D. Solar collectors and photovoltaics as combined heat and power systems: A critical review. *Energy Convers. Manag.* **2018**, *156*, 688–705. [CrossRef]

18. García-Segura, A.; Fernández-García, A.; Ariza, M.; Sutter, F.; Diamantino, T.; Martínez-Arcos, L.; Reche-Navarro, T.; Valenzuela, L. Influence of gaseous pollutants and their synergistic effects on the aging of reflector materials for concentrating solar thermal technologies. *Sol. Energy Mater. Sol. Cells* **2019**, *200*, 109955. [CrossRef]

19. Jin, J.; Ling, Y.; Hao, Y. Similarity analysis of parabolic-trough solar collectors. *Appl. Energy* **2017**, *204*, 958–965. [CrossRef]

20. Bellos, E.; Tzivanidis, C. Enhancing the performance of a parabolic trough collector with combined thermal and optical techniques. *Appl. Therm. Eng.* **2020**, *164*, 114496. [CrossRef]

21. Asociación Española para la Promoción de la Industria Termosolar. Informe de Transición del sector Eléctrico Horizonte 2030. 2018. Available online: http://www.protermosolar.com (accessed on 10 December 2019).

22. Bellos, E.; Tzivanidis, C. Alternative designs of parabolic trough solar collectors. *Prog. Energy Combust. Sci.* **2019**, *71*, 81–117. [CrossRef]

23. Bellos, E.; Tzivanidis, C. Investigation of a nanofluid-based concentrating thermal photovoltaic with a parabolic reflector. *Energy Convers. Manag.* **2019**, *180*, 171–182. [CrossRef]

24. Fernández-García, A.; Juaidi, A.; Sutter, F.; Martínez-Arcos, L.; Manzano-Agugliaro, F. Solar Reflector Materials Degradation Due to the Sand Deposited on the Backside Protective Paints. *Energies* **2018**, *11*, 808. [CrossRef]

25. Sánchez-Lozano, J.; García-Cascales, M.; Lamata, M. Evaluation of suitable locations for the installation of solar thermoelectric power plants. *Comput. Ind. Eng.* **2015**, *87*, 343–355. [CrossRef]

26. Kalogirou, S. *Solar Energy Engineering, Processes and Systems*; Academic Press: New York, NY, USA, 2014; 819p, ISBN 978-0-12-397270-5.

27. IDEA (Instituto para la Diversificación y Ahorro de la Energía). Anexo III. Cálculo de pérdidas de radiación solar por sombras. In *Instalaciones de Energía Solar Térmica, Pliego de Condiciones Técnicas de Instalaciones de Baja Temperatura*; IDEA: Madrid, Spain, 2011; pp. 40–44.

28. Castellano, N.N.; Parra, J.A.G.; Valls-Guirado, J.; Manzano-Agugliaro, F. Optimal displacement of photovoltaic array's rows using a novel shading model. *Appl. Energy* **2015**, *144*, 1–9. [CrossRef]

29. Fernández-García, A.; Zarza, E.; Valenzuela, L.; Perez, M. Parabolic-trough solar collectors and their applications. *Renew. Sustain. Energy Rev.* **2010**, *14*, 1695–1721. [CrossRef]

Article

Evolutionary Algorithms for Community Detection in Continental-Scale High-Voltage Transmission Grids

Manuel Guerrero [1], Raul Baños [2,*], Consolación Gil [1], Francisco G. Montoya [2] and Alfredo Alcayde [2]

[1] Dept. of Informatics, University of Almeria, E-04120 Almeria, Spain; mgl220@fm.ual.es (M.G.); cgilm@ual.es (C.G.)

[2] Dept. of Engineering, University of Almeria, E-04120 Almeria, Spain; pagilm@ual.es (F.G.M.); aalcayde@ual.es (A.A.)

* Correspondence: rbanos@ual.es; Tel.: +34-950014097

Received: 22 October 2019; Accepted: 28 November 2019; Published: 3 December 2019

Abstract: Symmetry is a key concept in the study of power systems, not only because the admittance and Jacobian matrices used in power flow analysis are symmetrical, but because some previous studies have shown that in some real-world power grids there are complex symmetries. In order to investigate the topological characteristics of power grids, this paper proposes the use of evolutionary algorithms for community detection using modularity density measures on networks representing supergrids in order to discover densely connected structures. Two evolutionary approaches (generational genetic algorithm, GGA+, and modularity and improved genetic algorithm, MIGA) were applied. The results obtained in two large networks representing supergrids (European grid and North American grid) provide insights on both the structure of the supergrid and the topological differences between different regions. Numerical and graphical results show how these evolutionary approaches clearly outperform to the well-known Louvain modularity method. In particular, the average value of modularity obtained by GGA+ in the European grid was 0.815, while an average of 0.827 was reached in the North American grid. These results outperform those obtained by MIGA and Louvain methods (0.801 and 0.766 in the European grid and 0.813 and 0.798 in the North American grid, respectively).

Keywords: power grids; supergrids; high-voltage power transmission; complex networks; community detection; modularity; evolutionary algorithms; generational genetic algorithm; modularity and improved genetic algorithm; Louvain modularity algorithm

---

1. Introduction

The optimal design and management of these supergrids is a difficult task, since it is necessary to manage large systems that include heterogeneous power grids from different countries. Most investigations in power systems often analyse optimisation problems such as optimal power flow, unit commitment, and economic dispatch, among others [1,2]. The solutions to these problems are often determined by the symmetry of the admittance and Jacobian matrices [3,4], and the topology of high-voltage transmission lines that connect the power produced at generating stations to substations, at which point the power flow is derived to other transmission lines or stepped down in voltage, and then submitted across power distribution lines into the end users. Many publications have addressed the factors that constrain the development of electricity infrastructure [5,6]. In particular, experts have highlighted that existing electric grids are inadequate to cope with increasing volumes of renewable electricity [7]. For example, the transmission systems in European countries are old, and a many miles of lines need to be replaced, upgraded, and even expanded to secure market integration, ensure supply security, and cope with the expansion in renewable energy planned for the next few years [8]. A similar challenge is faced in the United States, where renewable energy generation also accounts for an increasingly high percentage of annual demand [9].

Taking into account the fact that worldwide demand for electricity has been increasing and will continue to, it is necessary to ensure the reliable and secure operation of electricity transmission networks to efficiently transport energy from generation sources to electricity consumers. To achieve this goal, decisions need to be supported by expert systems able to process a large number of variables. Graph-based network analysis is a powerful tool for describing many real systems in a variety of fields [10]. Topological analysis provides the infrastructural information of power systems that is essential to assess network robustness or to generate synthetic power grids [11]. For example, some studies have detected complex symmetric subgraphs in large-scale power grids [12], and have provided a list of symmetric subgraphs with respect to reference nodes observed in the US grid.

Most real networks (graphs) representing real systems have clusters, such that many edges connect nodes within the same cluster, and comparatively few edges connect nodes in different clusters. This is why community detection [13–15] has gained popularity in recent years, especially among researchers working with complex networks [16–18]. In particular, community detection has been applied in field of electrical engineering, including the management of power grids [19–22]. However, keeping the complexity of the problem in mind, more work is needed to develop efficient algorithms to enable rapid community detection. With that aim, this paper evaluates the performance of evolutionary approaches for community detection in supergrids. These algorithms, which are guided by the modularity index [23] and consider different *degrees of abstraction* (i.e., detect any number of communities), enable a flexible and adaptive analysis of the power grid.

The remainder of the paper is organized as follows: Section 2 briefly describes the problem of community detection using graphs, and revises some previous studies that have been applied to electrical grids. Section 3 presents the main characteristics of two evolutionary algorithms used to solve the community detection problem using graphs [24]. Section 4 presents an empirical study that compares these methods for community detection in two supergrids. The conclusions of this work are provided in Section 5.

2. Community Detection

This section introduces the use of community detection in different research areas and discusses how community detection methods contribute to the analysis of power grids.

2.1. Community Detection: General Overview

Communities, also named clusters, are dense subgraphs which are well separated from each other. The community structure of complex networks reveals both their organisation and hidden relationships, among other elements [25]. In practice, a simple idea that has attained great popularity is that a community is a subgraph such that many edges connect nodes within the same group, and comparatively few edges connect nodes in different groups [14].

Many studies in different disciplines have shown that the community structure of complex networks reveals both their organisation and hidden relationships among their elements [25]. In particular, identifying communities can be useful for classifying the nodes in different groups [13]. So, nodes located at a central position in their community may have an important function of control and stability within the cluster, while those nodes located at the proximity of other communities can play a role of mediation or information exchange with these neighbouring communities.

An important consideration to be determined here is the number of communities to be detected. Some algorithms allow one to include a pre-established number of communities to be detected, while other approaches aim to infer the adequate number of communities depending of the characteristics of the networks [14]. A recently published survey paper [26] has reviewed a large number of community detection algorithms in multidisciplinary applications considering both disjoint and overlapping community detection problems. These applications include the study of social networks [25,27], communication networks [28,29], engineering systems and networks [18,30], biology and ecology [31,32], health sciences [33], scientometrics [34,35], economics [36], etc.

2.2. Community Detection in Power Grids

In recent years, the interest in the development of supergrids has grown remarkably. The supergrid concept was born as a solution to allow large-scale electrical power exchanges over continent-wide areas. This concept has been considered both a potential solution to transmission bottlenecks and an opportunity to trade higher volumes of electricity across longer distances [37]. In particular, we show the complexity of several high-voltage transmission topologies intended to connect two or more subsystems here, and note that supergrids have a meshed form to provide redundancy. In addition to the use of complex control methods [38], the variability of renewable sources [39] at continental scales can be mitigated by using the transmission grid and balancing locally with storage [40]. Some of the future major transmission projects around the world are described in [37]. For example, different projects aim to promote an efficient and reliable transmission grid in North America, including the *Tres Amigas superstation*. This superstation is the first version of this supergrid vision, since it is projected as a high-voltage direct current (HVDC) super-node asynchronously connecting the existing alternating current (AC) networks intended to link the three North American grids: the Eastern Interconnection, the Western Interconnection, and Texas Interconnection. This project involves a three-way alternating current/direct current (AC/DC) transmission superstation with several miles of underground superconducting DC cable, which will eliminate the market separation between the three asynchronous interconnections in the continental U.S. [41]. In the case of Europe, these authors indicate that an important number of major HVDC interconnections are being promoted to establish intercontinental interconnections with neighbouring regions with the aim of integrating regional energy markets into a single European market to achieve the European Union's (EU) renewable energy goals. Some authors have introduced the concept of global grid as the future stage of the electricity network, in which most of the large power plants in the world will be connected [42].

Some recent studies have proposed the analysis of the power grid infrastructure using graph-based network analysis techniques [19]. Usually, the nodes of the network represent the power plants and distribution and transmission substations, while the edges correspond to transmission lines. The application of graph-based analysis techniques has allowed for an analysis of the topological structure of networks representing power grids [43]. As commented above, a typical characteristic of all complex networks is the existence of community structures [13,15], such that detecting those communities can reveal the characteristics or functional relationships in a given network. In the case of power grids, communities represent substations densely connected by high-voltage transmission lines.

The importance of community detection in power grids comes from the fact that it is necessary to maintain grid reliability and enable more efficient restoration from severe disturbances. In particular, it is necessary to prepare a distribution grid for natural disasters (e.g., a storm), by developing switching plans to safely islands or disconnecting portions of the grid, preventing further degradation during incidents and enabling faster restoration after the disturbance. For example, reference [20] applied community detection to island power systems as an emergency response method to isolate failures that could propagate and lead to major disturbances. These authors developed two approaches based on modularity, with the DC power flow model incorporated into them, for islanding in medium and large networks and tested them in networks having 14, 30, 57, 118, and 247 nodes [20]. Other approaches use node similarity indexes to assign each node to the community sharing maximum similarity [22], and have demonstrated the good performance of this method in two IEEE standard power grids (39-bus standard power grid and 118-bus standard power grid). The IEEE 118-bus was also studied in [44]. Other researchers have presented a hierarchical spectral clustering method to reveal the internal connectivity of power transmission, establishing the possibility of islanding systems using a network with nodes and links representing buses and electrical transmission lines, respectively [21].

That approach was evaluated in several test systems of small, medium, and large sizes, including a model of Great Britain's transmission network [21]. Community detection has also been applied to analyse the vulnerability of the power systems under terrorist attacks [45], among other applications. However, none of these previous approaches have analysed supergrids.

3. Methodology

Two genetic algorithms designed to detect communities in graphs were applied in large networks modelling supergrids. These evolutionary approaches are guided by *modularity* [23], which is an applied objective function extensively used in community detection due to its simplicity and ease of calculation. Modularity provides a numerical value that represents the quality of the solution, with greater values corresponding to a more accurate community structure. Therefore, the aim was to find communities that maximise the value of *modularity* (*Q*), defined as:

$$Q = \frac{1}{2M} \sum \left(a_{ij} - \frac{K_i K_j}{2M} \right) \delta(i,j), \tag{1}$$

where M represents the total number of edges in the network; the sub-indices i and j indicate two nodes (vertices) of the network, K_i and K_j being the degrees of the i-th and j-th nodes, respectively; the parameter a_{ij} is the element of the i-th row and the j-th column of the adjacency matrix; and $\delta(i,j)$ represents the relationship between the i-th node and the j-th node (i.e., $\delta(i,j) = 1$ if node i and node j are in the same community; otherwise, $\delta(i,j) = 0$).

Finding these communities by maximising the modularity [23] or another objective function is an NP-hard problem [46]. Brandes et al. [47] proved that modularity maximisation is an NP-hard problem, even for the restricted version with a bond of two clusters, and suggested further investigation of approximation algorithms and heuristics for solving this problem. More recently, other authors have demonstrated the high complexity of calculating modularity on sparse graphs and dense graphs separately [48]. Due to the high complexity of the community detection problem, researchers have applied heuristic and meta-heuristic methods to obtain high quality solutions, in a reasonable computational time.

The field of evolutionary computation [49] is closely related to computational intelligence, with a focus on designing algorithms to solve complex global optimisation problems. Evolutionary algorithms are problem-solving procedures that include evolutionary processes as the key design elements, such that a population of individuals is continually and selectively evolved until a termination criteria is fulfilled. Genetic algorithms (GAs) [50] are probably the most widely used evolutionary techniques. A genetic algorithm mimics natural selection by evolving, over time, a population of individual solutions to the problem at hand until a termination condition is fulfilled and the best individual is taken as an acceptable solution. Two important characteristics of GAs are the representation used (e.g., binary or real) and the genetic operators employed (e.g., mutation and crossover). GAs have been successfully applied to solve electrical problems [51–53].

In this study, two genetic algorithms were adapted to solve community detection problems in power grids. These algorithms (MIGA and GGA+) were recently proposed and shown to be more effective than other approaches to community detection, as assessed by benchmarks typically used to compare algorithms. Figure 1 shows the flowchart of MIGA and GGA+. The main characteristics of both methods are briefly described below.

- The modularity and improved genetic algorithm (MIGA) [54] takes the modularity (*Q*) as the objective function, and uses the number of community structures as prior information to improve stability and accuracy of community detection. MIGA also uses simulated annealing [55] as local search strategy. Note that many authors have previously considered the use of local search strategies [56] for solving hard optimisation problems.

- The generational genetic algorithm (GGA+): GGA+ [57] includes efficient and safe initialisation methods in which a maximum node size is assigned to each community. Several operators are applied to migrate or exchange nodes between communities while using the modularity function as the objective function. An important feature of GGA+ is that it is able to rapidly obtain community partitions with different degrees of abstraction.

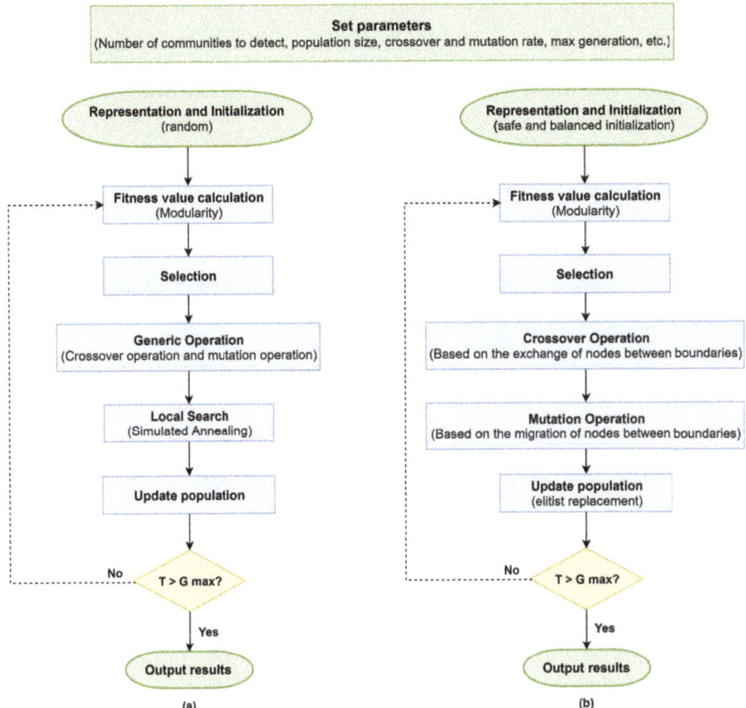

Figure 1. Flowchart of (**a**) MIGA and (**b**) GGA+.

4. Empirical Study

This section analyses the performance of the MIGA and GGA+ algorithms in detecting communities in two networks representing supergrids. Neglecting complex electrical properties, the nodes of the network represent the power plants and distribution and transmission substations, while the edges correspond to transmission lines. In this way, the power grid is simplified as an undirected and unweighted network.

4.1. Test Cases

To analyse the performance of the genetic algorithms, graphical models of two supergrids were considered: the European grid, including part of Russia, North Africa, and part of the Near East; and the North American grid.

- The European electric network was obtained from the European Network of Transmission System Operators (ENTSO-E) [58]. The ENTSO-E group consists of 43 electricity transmission system operators from 36 countries across Europe who are responsible for the bulk transmission of electric power on the main high-voltage electric networks. This power network, which also includes data from North African and Near Eastern countries, is formed by transmission lines designed for 220 kV voltage and higher and generation stations with a net generation capacity of more than 100 MW.

- The North American electric network was obtained using the GridKit 1.0 toolkit, which was developed in the context of the SciGRID project at the NEXT ENERGY-EWE Research Centre for Energy Technology [59]. GridKit is a power grid extraction tool that converts geographical objects representing elements of power systems in OpenStreetMap to model the electric network. This network covers the United States, Canada, and Mexico and includes transmission lines that operate at relatively high-voltages varying from 69 kV up to 765 kV. The power grid of the United States is probably the best system studied in the literature, due to the particular characteristics of the network. The regions covered receive their bulk electricity from three separate electric grids: the huge Eastern Interconnection, the Western Interconnection, and the relatively small Texas grid [60], which is almost entirely managed by the Electric Reliability Council of Texas (ERCOT).

Table 1 describes some graphical characteristics of these networks. The number of nodes and edges is very large, which denotes the high complexity of community detection in these networks. Note that the dimensions of these networks are significantly larger than other power grids considered in recent studies (see, e.g., [22]). In fact, to our knowledge, no previous paper has applied graph-based analytical methods to power networks of these dimensions.

Table 1. Graph description of the European (EU) and North American (NA) power grids.

Feature	EU Grid	NA Grid
Nodes	7893	16,063
Edges	10,346	20,169
Average degree	2.62	2.51
Network diameter	108	158
Avg. path length	41.27	48.99
Avg. clustering coefficient	0.07	0.01
Eigenvector centrality	1.70	1.87

4.2. Parameter Configuration

To perform a fair comparison between the two evolutionary algorithms, the parameters were set to the values established in the original publications. The influence of GGA+ parameters was adjusted by means of a sensitivity analysis method based on executing the algorithms with different number of individuals and probabilities of using evolutionary search operators [44]. Statistical results obtained from these independent runs were considered to select the following parameters: the population size was set to 200 individuals; the number of iterations (generations) of the algorithm was set to 200; and the probability of applying the search operators was set between 20% and 80%. As commented above, MIGA also uses simulated annealing [55] as a local search strategy, with the following parameters: initial temperature 800,000, cooling rate 0.99, and minimum temperature 0.01. The experiments were performed on a personal computer with an Intel Core i7 3630Q processor (2.4 GHz, 8 GB DDR3 RAM), which executed the application we developed in the C# .Net Framework 4.

4.3. Results and Discussion

The accuracies of MIGA and GGA+ were evaluated according to the Q values. Table 2 shows the maxima, means, minima, and standard deviations (SD) of the modularities obtained by MIGA and GGA+ in the European (EA) and the North American (NA) grids considering c = {2, 3, 4, 5, 10, 20, 30, 40, 50} communities. A number of communities within the range of 2 to 50 were used to show how evolutionary algorithms are able to work under different levels of abstraction. However, these algorithms could be applied to obtain a greater number of communities, although their size would decrease considerably. The median runtimes (in minutes) of these 30 independent runs are also provided. Furthermore, the communities detected by Louvain modularity method implemented in

Gephi are also shown in this table. To conduct the performance analysis and to statistically compare the quality of the solutions obtained by the two algorithms, a total of 30 independent runs were performed with each algorithm on each network.

Table 2. Results obtained by MIGA and GGA+ after 30 independent runs and comparison with Louvain method implemented in Gephi (runtime in minutes).

Grid	Method	Metric	$c = 2$	$c = 3$	$c = 4$	$c = 5$	$c = 10$	$c = 20$	$c = 30$	$c = 40$	$c = 50$	AVG
EU	MIGA	Max(Q)	0.496	0.660	0.740	0.785	0.877	0.911	0.921	0.924	0.922	0.804
		Mean(Q)	0.495	0.654	0.736	0.781	0.872	0.909	0.918	0.922	0.921	0.801
		Min(Q)	0.491	0.612	0.730	0.777	0.866	0.907	0.916	0.920	0.919	0.793
		SD(Q)	0.001	0.009	0.002	0.002	0.002	0.001	0.001	0.001	0.001	0.002
		Mean time	515	632	404	409	289	296	205	237	246	359
	GGA+	Max(Q)	0.498	0.665	0.746	0.793	0.889	0.929	0.941	0.943	0.948	0.817
		Mean(Q)	0.498	0.663	0.744	0.793	0.887	0.927	0.938	0.942	0.947	0.815
		Min(Q)	0.496	0.661	0.742	0.791	0.885	0.925	0.937	0.940	0.946	0.814
		SD(Q)	0.000	0.001	0.001	0.001	0.001	0.001	0.001	0.001	0.000	0.001
		Mean time	661	614	556	417	443	462	314	228	244	438
	Louvain	(Q)	0.291	0.599	0.677	0.699	0.874	0.929	0.939	0.944	0.945	0.766
NA	MIGA	Max(Q)	0.498	0.663	0.743	0.788	0.882	0.924	0.939	0.945	0.948	0.814
		Mean(Q)	0.498	0.659	0.740	0.789	0.878	0.922	0.937	0.944	0.947	0.813
		Min(Q)	0.497	0.656	0.736	0.785	0.872	0.921	0.936	0.943	0.946	0.810
		SD(Q)	0.000	0.002	0.002	0.002	0.003	0.001	0.001	0.000	0.000	0.001
		Mean time	1228	770	708	747	425	317	423	559	244	602
	GGA+	Max(Q)	0.499	0.670	0.753	0.804	0.900	0.943	0.956	0.963	0.967	0.828
		Mean(Q)	0.499	0.669	0.753	0.802	0.898	0.940	0.955	0.962	0.966	0.827
		Min(Q)	0.498	0.667	0.751	0.801	0.893	0.936	0.954	0.961	0.965	0.825
		SD(Q)	0.000	0.001	0.001	0.001	0.001	0.002	0.001	0.000	0.000	0.001
		Mean time	1909	1335	1221	568	521	764	593	552	298	862
	Louvain	(Q)	0.478	0.608	0.652	0.754	0.881	0.935	0.952	0.959	0.963	0.798

These results show that GGA+ achieved the best mean and maximum values in both grids, regardless of the number of communities to be detected. These results also indicate that the greater the number of communities, the greater the advantage of GGA+ over MIGA. In addition, the standard deviation obtained from the results of these 30 independent runs was often smaller for GGA+ than for MIGA. The modularity values obtained by both algorithms increased with the number of communities without degradation of the standard deviation, indicating the robustness of these evolutionary approaches.

Table 2 shows that the runtime required by both evolutionary algorithms is of the same order of magnitude in the North American and European networks, while the differences come from the fact that the former has approximately double the number of nodes and edges as the latter (see Table 1). On the other hand, in both cases the runtime tends to decrease when the number of communities is greater. This is due to the crossover and mutation operators moving a given percentage of the nodes between a community and a neighbouring one. Therefore, the bigger the communities are, the higher that number of nodes that are moved between neighbouring communities, and therefore, the runtime increases. It can be concluded that GGA+ is scalable both in terms of network size and in terms of number of communities. Both algorithms require a few hours to complete the search process with these parameter settings, which is not a critical issue since the goal is to find solutions with greater modularity regardless of the execution time. Of course, the execution time could be reduced considerably by modifying the parameter settings or applying parallel processing techniques.

When two algorithms are compared, it is common to determine whether there are significant differences between the solutions they obtain. With this aim, a one-way ANOVA was applied, with the results indicating that the p-value was <0.05 in all cases; i.e., the null hypothesis was always rejected, which means that there was a significant difference between at least some of the means of the different groups. Thus, the results obtained by GGA+ were significantly different from those obtained by MIGA, validating the mean values in Table 2.

The analysis of Figure 2 reveals that there are some differences between the results obtained by the two methods when detecting three and twenty communities, especially when the number of communities increases. Each community in these networks is represented with a random colour. Even in the case of detecting only three communities, MIGA has some difficulties in assigning communities in some parts of the graph, while GGA+ obtains clearly differentiated communities. Considering these graphical results and the results provided in Table 2, it can be concluded that GGA+ not only outperforms to MIGA, but also exhibits good performance in these large networks.

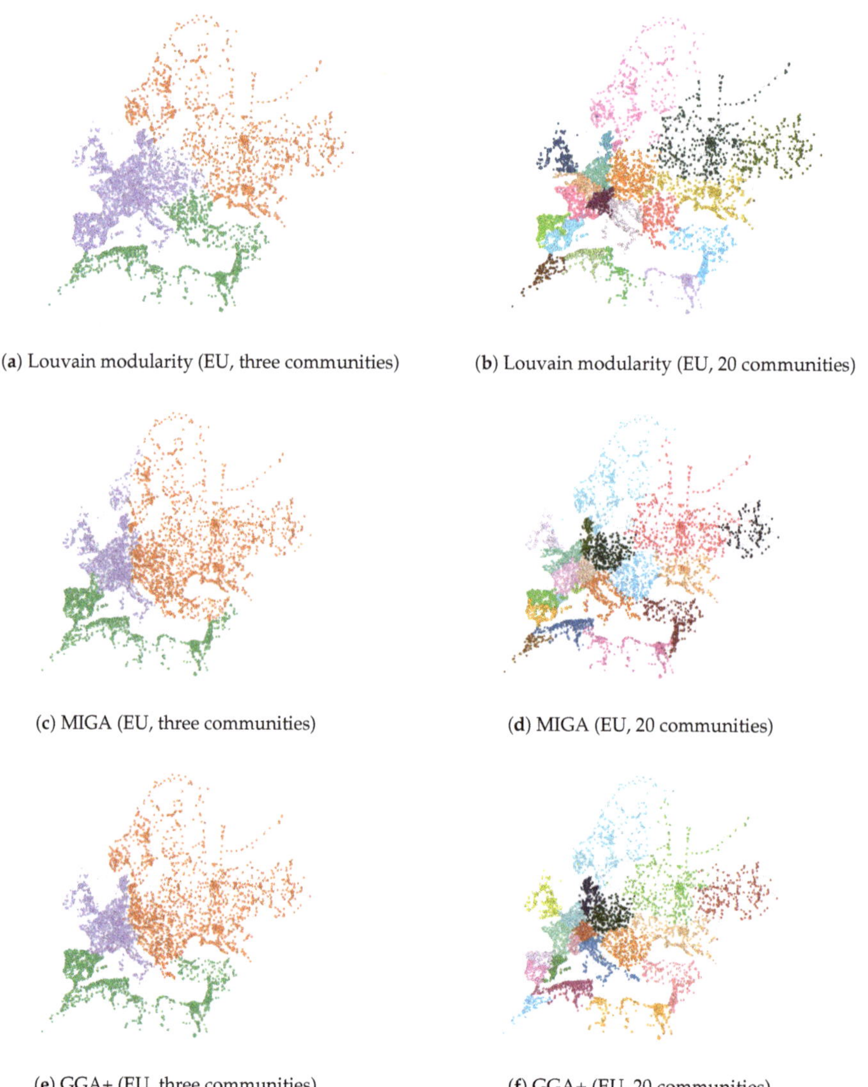

(**a**) Louvain modularity (EU, three communities)	(**b**) Louvain modularity (EU, 20 communities)
(**c**) MIGA (EU, three communities)	(**d**) MIGA (EU, 20 communities)
(**e**) GGA+ (EU, three communities)	(**f**) GGA+ (EU, 20 communities)

Figure 2. Results obtained by Louvain modularity method, MIGA, and GGA+ for the European grid (three and 20 communities).

The results obtained by GGA+ are analysed in more detail here. Figure 3a–c display the communities detected by GGA+ in the European power grid with 5, 10, and 30 communities. These results reveal that this algorithm is able to obtain good quality solutions even when the number of communities increases. Moreover, Figure 3d–f provide a different layout based on the ForceAtlas2 [61] plugin in Gephi for these three networks. While the results presented in Figure 3a–c correspond to the coordinates of each node, the results displayed in Figure 3d–f cannot be read as a Cartesian projection. Instead, ForceAtlas2 was in a drawing mode that has the specificity of placing each node depending on the other nodes. This visualisation method builds a force directed layout by simulating a physical system in order to accommodate nodes and links in a spatial network. Nodes repel each other like charged particles, while edges attract their nodes like springs. Note that the same colour is used to represent the physical layout and the distribution obtained by ForceAtlas2. Moreover, the number of nodes in each community is often balanced (e.g., the five communities obtained in the European grid have a percentage of nodes between 19.16% and 20.47% of the total of nodes), although there are some significant imbalances between clusters when the number of communities increases (e.g., 30 communities). The analysis of Figure 3 demonstrates the good behaviour of GGA+ independent of the degree of abstraction. Figure 4 shows how geographical structures change with the number of communities.

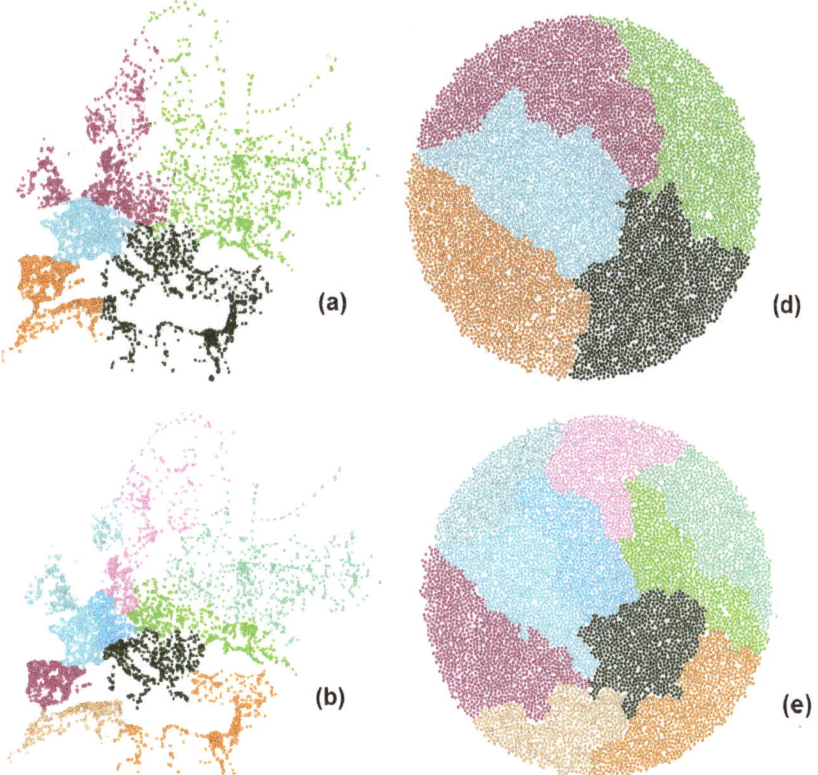

Figure 3. *Cont.*

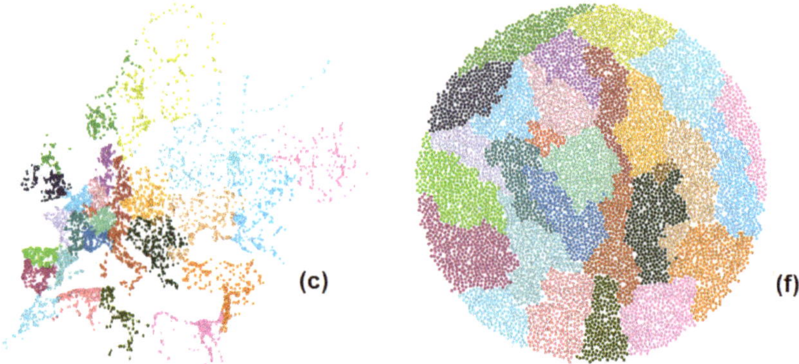

Figure 3. Results obtained by GGA+ for the European power grid: physical layout with (**a**) five communities, (**b**) 10 communities, and (**c**) 30 communities. Distribution obtained by ForceAtlas2 with (**d**) five communities, (**e**) 10 communities, and (**f**) 30 communities.

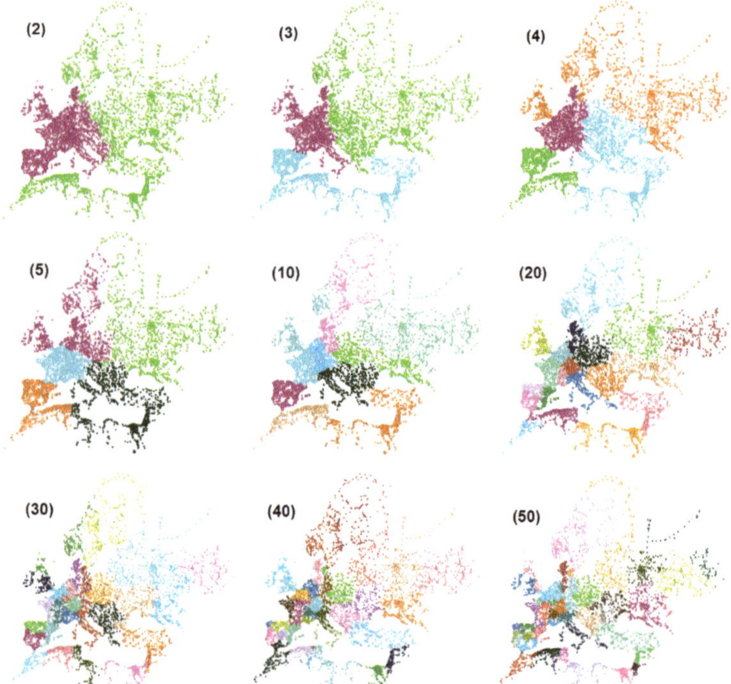

Figure 4. Physical layout of the communities detected by GGA+ in the European network using different degrees of abstraction (the number of communities is indicated in parentheses).

The analysis of the North American network supports similar conclusions. Thus, Figure 5a–c display the results obtained by GGA+ in that network when 5, 10, and 30 communities are detected. These data reveal that this algorithm is able to obtain good quality solutions not only with a few communities, but when the number of communities increases. The results obtained by the layout provided by ForceAtlas2 for these configurations (Figure 5d–f) demonstrate the good behaviour of

GGA+ independent of the degree of abstraction. Finally, Figure 5 shows that the algorithm is able to obtain differentiated clusters, even when the number of communities increases significantly. The results obtained here are of particular interest, bearing in mind that the North American electrical grid is made up of three interconnections: the Western Interconnection, the Eastern Interconnection, and the ERCOT (Texas) Interconnection, which are not synchronised, and alternating current (AC) power must be converted to direct current (DC) power for transfer across any of the interconnections. To overcome these limitations, the *Tres Amigas* superstation has been planned in New Mexico (U.S.), a 1.6 billion dollar project that aims to connect these three primary interconnections to facilitate the smooth, reliable, and efficient transfer of green power from region to region while integrating substantial renewable energy sources [62]. Figure 6 shows how geographical structures in the North American grid change with the number of communities.

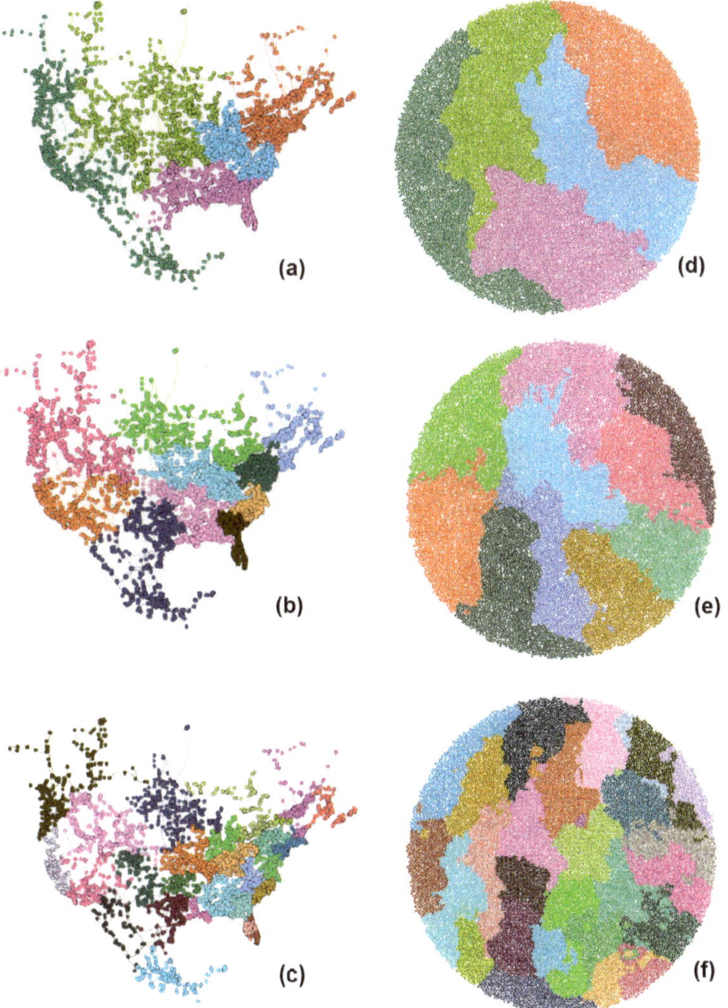

Figure 5. Results obtained by GGA+ for the North American power grid: physical layout with (**a**) five communities, (**b**) 10 communities, and (**c**) 30 communities. Distribution obtained by ForceAtlas2 with (**d**) five communities, (**e**) 10 communities, and (**f**) 30 communities.

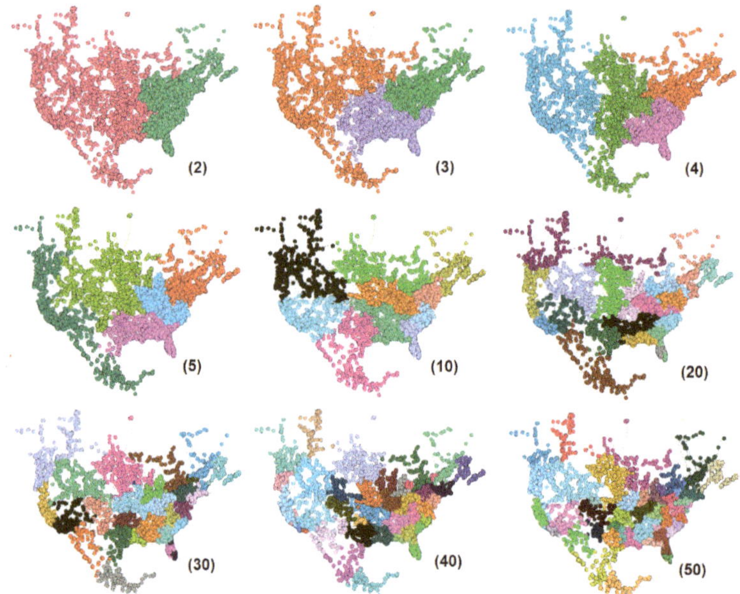

Figure 6. Physical layout of the communities detected by GGA+ in the North American network using different degrees of abstraction (the number of communities is indicated in parentheses).

5. Conclusions

The optimal design of high-voltage transmission networks is a critical issue to supply electrical energy to residential areas and industries. In fact, the growing integration of power grids across regions requires investment in more transmission power supply systems to ensure system stability and guarantee power supplies. To reach that aim, it is important to investigate the topological characteristics of these supergrids. This paper opens a new avenue of research by analysing the community structures in supergrids in a fast and effective way. In particular, it is shown that solving the community detection problem with evolutionary algorithms allows one to obtain some key ideas about the structure of these networks. In particular, two evolutionary methods that include powerful initialisation methods and evolutionary search operators under the guidance of modularity were used to detect communities in large-scale networks. The evolutionary algorithms adopted a flexible and adaptive analysis of the characteristics of the power grids with different levels of detail (number of communities). The empirical study considered two large networks representing supergrids: (i) Europe, including Russia, North Africa, and part of the Near East (7893 nodes and 10,346 branches); and (ii) North America (16,063 nodes and 20,169 edges). In particular, these methods were able to partition the networks into some loosely coupled sub-networks (communities) of similar scale, such that nodes within a community were densely linked, while connections between different communities were sparser. Numerical and graphical analysis using graph visualisation tools showed that GGA+ slightly outperformed MIGA, especially when the number of communities increased. Both evolutionary approaches outperformed the modularity values of the communities detected by the Louvain method implemented in Gephi. The results obtained show that evolutionary approaches are efficient methods for detecting communities in supergrids having thousand of nodes, and provide interesting topological information about the physical distribution and concentration of these elements of the grids. Future work will apply parallel and multi-objective optimisation methods and include the electrical properties of the power networks.

Author Contributions: Conceptualisation, M.G. and R.B.; methodology, C.G. and R.B.; software, M.G. and R.B.; validation, M.G., R.B., and A.A.; formal analysis, F.G.M. and C.G.; investigation, M.G., R.B., and C.G.; resources, F.G.M.; data curation, R.B. and A.A.; writing—original draft preparation, R.B., F.G.M., and A.A.; writing—review and editing, R.B., F.G.M., and C.G.; visualisation, A.A. and F.G.M.; supervision, C.G. and R.B.; project administration, C.G. and F.G.M.

Funding: This research received funding by Spanish Ministry of Science, Innovation, and Universities (project PGC2018-098813-B-C33).

Acknowledgments: This research was supported by the Spanish Ministry of Science, Innovation and Universities (project PGC2018-098813-B-C33), and by the Regional Government of Andalusia (ceiA3 project).

Conflicts of Interest: The authors declare no conflict of interest.

References

1. Alqurashi, A.; Etemadi, A.H.; Khodaei, A. Treatment of uncertainty for next generation power systems: State-of-the-art in stochastic optimization. *Electr. Power Syst. Res.* **2016**, *141*, 233–245. [CrossRef]

2. Wang, Y.; Wang, S.; Wu, L. Distributed optimization approaches for emerging power systems operation: A review. *Electr. Power Syst. Res.* **2017**, *144*, 127–135. [CrossRef]

3. Lin, W.M.; Teng, J.H. Three-phase distribution network fast-decoupled power flow solutions. *Int. J. Electr. Power Energy Syst.* **2000**, *22*, 375–380. [CrossRef]

4. Kamel, S.; Jurado, F. Power flow analysis with easy modelling of interline power flow controller. *Electr. Power Syst. Res.* **2014**, *108*, 234–244. [CrossRef]

5. Battaglini, A.; Komendantova, N.; Brtnik, P.; Patt, A. Perception of barriers for expansion of electricity grids in the European Union. *Energy Policy* **2012**, *47*, 254–259. [CrossRef]

6. Codetta-Raiteri, D.; Bobbio, A.; Montani, S.; Portinale, L. A dynamic Bayesian network based framework to evaluate cascading effects in a power grid. *Eng. Appl. Artif. Intell.* **2012**, *25*, 683–697. [CrossRef]

7. IEA; OECD. Security of Supply in Electricity Markets. Evidence and Policy Issues. 2002. Available online: https://www.oecd-ilibrary.org/energy/security-of-supply-in-electricity-markets_9789264174504-en (accessed on 11 November 2019).

8. ENTSO-E. Ten-Year Network Development Plan 2010–2020. 2010. Available online: https://www.entsoe.eu/fileadmin/user_upload/_library/SDC/TYNDP/TYNDP-final_document.pdf (accessed on 11 November 2019).

9. Cochran, J.; Denholm, P.; Speer, B.; Miller, M. *Grid Integration and the Carrying Capacity of the U.S. Grid to Incorporate Variable Renewable Energy*; Technical Report NREL/TP-6A20-62607; National Renewable Energy Laboratory: Golden, CO, USA, 2015.

10. Bornholdt, S.; Schuster, H.G. *Handbook of Graphs and Networks: From the Genome to the Internet*; John Wiley & Sons, Inc.: New York, NY, USA, 2003.

11. Espejo, R.; Lumbreras, S.; Ramos, A. Analysis of transmission-power-grid topology and scalability, the European case study. *Phys. A Stat. Mech. Appl.* **2018**, *509*, 383–395. [CrossRef]

12. MacArthur, B.D.; Sánchez-García, R.J.; Anderson, J.W. Symmetry in complex networks. *Discret. Appl. Math.* **2008**, *156*, 3525–3531. [CrossRef]

13. Fortunato, S. Community detection in graphs. *Phys. Rep.* **2010**, *486*, 75–174. [CrossRef]

14. Fortunato, S.; Hric, D. Community detection in networks: A user guide. *Phys. Rep.* **2016**, *659*, 1–44. [CrossRef]

15. Newman, M.E. Communities, modules and large-scale structure in networks. *Nat. Phys.* **2012**, *8*, 25–31. [CrossRef]

16. Xiao, J.; Li, X.; Chen, S.; Wang, Y.; Han, J.; Zhou, Z. Complex network measurement and optimization of Chinese domestic movies with internet of things technology. *Comput. Electr. Eng.* **2017**. [CrossRef]

17. Xiao, Z.; Cao, B.; Zhou, G.; Sun, J. The monitoring and research of unstable locations in eco-industrial networks. *Comput. Ind. Eng.* **2017**, *105*, 234–246. [CrossRef]

18. He, X.; Dong, Y.; Wu, Y.; Wei, G.; Xing, L.; Yan, J. Structure analysis and core community detection of embodied resources networks among regional industries. *Phys. A Stat. Mech. Appl.* **2017**, *479*, 137–150. [CrossRef]

19. Pagani, G.A.; Aiello, M. The Power Grid as a complex network: A survey. *Phys. A Stat. Mech. Appl.* **2013**, *392*, 2688–2700. [CrossRef]

20. Pahwa, S.; Youssef, M.; Schumm, P.; Scoglio, C.; Schulz, N. Optimal intentional islanding to enhance the robustness of power grid networks. *Phys. A Stat. Mech. Appl.* **2013**, *392*, 3741–3754. [CrossRef]

21. Sánchez-García, R.J.; Fennelly, M.; Norris, S.; Wright, N.; Niblo, G.; Brodzki, J.; Bialek, J.W. Hierarchical Spectral Clustering of Power Grids. *IEEE Trans. Power Syst.* **2014**, *29*, 2229–2237. [CrossRef]

22. Chen, Z.; Xie, Z.; Zhang, Q. Community detection based on local topological information and its application in power grid. *Neurocomputing* **2015**, *170*, 384–392. [CrossRef]

23. Newman, M.E.; Girvan, M. Finding and evaluating community structure in networks. *Phys. Rev. E* **2004**, *69*, 026113. [CrossRef]

24. IEA. Large-Scale Electricity Interconnection. Technology and Prospects for Cross-Regional Networks. 2016. Available online: https://www.iea.org/publications/freepublications/publication/Interconnection.pdf (accessed on 11 November 2019).

25. Reihanian, A.; Feizi-Derakhshi, M.R.; Aghdasi, H.S. Community detection in social networks with node attributes based on multi-objective biogeography based optimization. *Eng. Appl. Artif. Intell.* **2017**, *62*, 51–67. [CrossRef]

26. Javed, M.A.; Younis, M.S.; Latif, S.; Qadir, J.; Baig, A. Community detection in networks: A multidisciplinary review. *J. Netw. Comput. Appl.* **2018**, *108*, 87–111. [CrossRef]

27. Kanavos, A.; Perikos, I.; Hatzilygeroudis, I.; Tsakalidis, A. Emotional community detection in social networks. *Comput. Electr. Eng.* **2018**, *65*, 449–460. [CrossRef]

28. Nguyen, N.P.; Dinh, T.N.; Shen, Y.; Thai, M.T. Dynamic social community detection and its applications. *PLoS ONE* **2014**, *9*, e91431. [CrossRef] [PubMed]

29. Soliman, S.S.; El-Sayed, M.F.; Hassan, Y.F. Semantic clustering of search engine results. *Sci. World J.* **2015**, *2015*, 931258. [CrossRef] [PubMed]

30. Kicsi, A.; Csuvik, V.; Vidács, L.; Horváth, F.; Beszédes, Á.; Gyimóthy, T.; Kocsis, F. Feature analysis using information retrieval, community detection and structural analysis methods in product line adoption. *J. Syst. Softw.* **2019**, *155*, 70–90. [CrossRef]

31. Atay, Y.; Koc, I.; Babaoglu, I.; Kodaz, H. Community detection from biological and social networks: A comparative analysis of metaheuristic algorithms. *Appl. Soft Comput.* **2017**, *50*, 194–211. [CrossRef]

32. Rozylowicz, L.; Nita, A.; Manolache, S.; Popescu, V.D.; Hartel, T. Navigating protected areas networks for improving diffusion of conservation practices. *J. Environ. Manag.* **2019**, *230*, 413–421. [CrossRef]

33. Soofi, S.; Ahmed, S.; Fox, M.P.; MacLeod, W.B.; Thea, D.M.; Qazi, S.A.; Bhutta, Z.A. Effectiveness of community case management of severe pneumonia with oral amoxicillin in children aged 2–59 months in Matiari district, rural Pakistan: A cluster-randomised controlled trial. *Lancet* **2012**, *379*, 729–737. [CrossRef]

34. Gómez-Núñez, A.J.; Batagelj, V.; Vargas-Quesada, B.; Moya-Anegón, F.; Chinchilla-Rodríguez, Z. Optimizing SCImago Journal & Country Rank classification by community detection. *J. Informetr.* **2014**, *8*, 369–383.

35. Carusi, C.; Bianchi, G. Scientific community detection via bipartite scholar/journal graph co-clustering. *J. Informetr.* **2019**, *13*, 354–386. [CrossRef]

36. Jackson, M.O. An overview of social networks and economic applications. In *Handbook of Social Economics*; North-Holland (Elsevier): Amsterdam, The Netherlands, 2011; Volume 1, pp. 511–585.

37. Arcia-Garibaldi, G.; Cruz-Romero, P.; Gómez-Expósito, A. Future power transmission: Visions, technologies and challenges. *Renew. Sustain. Energy Rev.* **2018**, *94*, 285–301. [CrossRef]

38. DeMarco, C.L.; Baone, C.A. Control of Power Systems with High Penetration Variable Generation. In *Renewable Energy Integration: Practical Management of Variability, Uncertainty, and Flexibility in Power Grids*, 2nd ed.; Academic Press (Elsevier): San Diego, CA, USA, 2017; pp. 385–394.

39. Wu, H.; Krad, I.; Florita, A.; Hodge, B.M.; Ibanez, E.; Zhang, J.; Ela, E. Stochastic multi-timescale power system operations with variable wind generation. *IEEE Trans. Power Syst.* **2016**, *32*, 3325–3337. [CrossRef]

40. Schlachtberger, D.; Brown, T.; Schramm, S.; Greiner, M. The benefits of cooperation in a highly renewable European electricity network. *Energy* **2017**, *134*, 469–481. [CrossRef]

41. Teichler, S.L.; Levitine, I. HVDC transmission: A path to the future? *Electr. J.* **2010**, *23*, 27–41. [CrossRef]

42. Chatzivasileiadis, S.; Ernst, D.; Andersson, G. The global grid. *Renew. Energy* **2013**, *57*, 372–383. [CrossRef]

43. Barthelemy, M. Spatial Networks. *Phys. Rep.* **2011**, *499*, 1–101. [CrossRef]

44. Guerrero, M.; Montoya, F.G.; Baños, R.; Alcayde, A.; Gil, C. Community detection in national-scale high voltage transmission networks using genetic algorithms. *Adv. Eng. Inform.* **2018**, *38*, 232–241. [CrossRef]

45. Wang, S.; Zhang, J.; Zhao, M.; Min, X. Vulnerability analysis and critical areas identification of the power systems under terrorist attacks. *Phys. A Stat. Mech. Appl.* **2017**, *473*, 156–165. [CrossRef]

46. Nascimento, M.C.; Pitsoulis, L. Community detection by modularity maximization using GRASP with path relinking. *Comput. Oper. Res.* **2013**, *40*, 3121–3131. [CrossRef]

47. Brandes, U.; Delling, D.; Gaertler, M.; Gorke, R.; Hoefer, M.; Nikoloski, Z.; Wagner, D. On modularity clustering. *IEEE Trans. Knowl. Data Eng.* **2008**, *20*, 172–188. [CrossRef]

48. DasGupta, B.; Desai, D. On the complexity of Newman's community finding approach for biological and social networks. *J. Comput. Syst. Sci.* **2013**, *79*, 50–67. [CrossRef]

49. Fogel, D.B. *Evolutionary Computation: Toward a New Philosophy of Machine Intelligence*, 3rd ed.; John Wiley & Sons, Inc.: Hoboken, NJ, USA, 2006; 384p.

50. Holland, J.H. *Adaptation in Natural and Artificial Systems: An Introductory Analysis with Applications to Biology, Control, and Artificial Intelligence*; University of Michigan Press: Ann Arbor, MI, USA, 1975; 183p.

51. Sánchez, P.; Montoya, F.G.; Manzano-Agugliaro, F.; Gil, C. Genetic algorithm for S-transform optimisation in the analysis and classification of electrical signal perturbations. *Expert Syst. Appl.* **2013**, *40*, 6766–6777. [CrossRef]

52. Wu, Y.; Tang, Y.; Han, B.; Ni, M. A topology analysis and genetic algorithm combined approach for power network intentional islanding. *Int. J. Electr. Power Energy Syst.* **2015**, *71*, 174–183. [CrossRef]

53. Thakur, M.; Kumar, A. Optimal coordination of directional over current relays using a modified real coded genetic algorithm: A comparative study. *Int. J. Electr. Power Energy Syst.* **2016**, *82*, 484–495. [CrossRef]

54. Shang, R.; Bai, J.; Jiao, L.; Jinm, C. Community detection based on modularity and an improved genetic algorithm. *Phys. A Stat. Mech. Appl.* **2013**, *392*, 1215–1231. [CrossRef]

55. Kirkpatrick, S.; Vecchi, M.P.; Gelatt, C.D. Optimization by simulated annealing. *Science* **1983**, *220*, 671–680. [CrossRef]

56. Kizys, R.; Juan, A.A.; Sawik, B.; Calvet, L. A biased-randomized iterated local search algorithm for rich portfolio optimization. *Appl. Sci.* **2019**, *9*, 3509. [CrossRef]

57. Guerrero, M.; Montoya, F.G.; Baños, R.; Alcayde, A.; Gil, C. Adaptive community detection in complex networks using genetic algorithms. *Neurocomputing* **2017**, *266*, 101–113. [CrossRef]

58. ENTSO-E. ENTSO-E Transmission System Map. 2019. Available online: https://www.entsoe.eu/map/Pages/default.aspx (accessed on 11 November 2019).

59. Wiegmans, B. GridKit 1.0 'for Scientists' (Version v1.0). Zenodo. 2016. Available online: https://zenodo.org/record/47263#.XeT6e2ZumiM (accessed on 11 November 2019).

60. Mazur, A.; Metcalfe, T. America's three electric grids: Are efficiency and reliability functions of grid size? *Electr. Power Syst. Res.* **2012**, *89*, 191–195. [CrossRef]

61. Jacomy, M.; Venturini, T.; Heymann, S.; Bastian, M. ForceAtlas2, a continuous graph layout algorithm for handy network visualization designed for the Gephi software. *PLoS ONE* **2014**, *9*, e98679. [CrossRef]

62. TresAmigas. Tres Amigas Superstation Project. 2019. Available online: http://www.tresamigasllc.com/ (accessed on 11 November 2019).

Article

Geometric Algebra in Nonsinusoidal Power Systems: A Case of Study for Passive Compensation

Francisco G. Montoya

Department of Engineering, University of Almeria, CEIA3, 04120 Almeria, Spain; pagilm@ual.es

Received: 20 September 2019; Accepted: 11 October 2019; Published: 14 October 2019

Abstract: New-generation power networks, such as microgrids, are being affected by the proliferation of nonlinear electronic systems, resulting in harmonic disturbances both in voltage and current that affect the symmetry of the system. This paper presents a method based on the application of geometric algebra (GA) to the resolution of power flow in nonsinusoidal single-phase electrical systems for the correct determination of its components to achieve passive compensation of true quadrature current. It is demonstrated that traditional techniques based on the concepts of Budeanu, Fryze or IEEE1459 fail to determine the interaction between voltage and current and therefore, are not suitable for being used as a basis for the compensation of nonactive power components. An example is included that demonstrates the superiority of GA method and is compared to previous work where GA approaches and traditional methods have also been used.

Keywords: geometric algebra; nonsinusoidal power; passive compensation; clifford algebra; circuit systems

---

1. Introduction

The new power grids are a major step forward for today's society, as they allow better energy management and integration with new renewable sources such as solar, wind, etc. [1]. These networks are made up of a large number of devices based on power electronics. A clear example is seen in distributed generation systems, intelligent buildings or control systems, where many receivers are installed such as cycloconverters, speed drives, household appliances, battery power converters, power inverters and more. Likewise, symmetry is a fundamental concept in art as well as science and engineering. Although, during the normal operation of the network, the system usually presents a symmetry in the waveform of voltage and current, all these elements can cause the network to supply a highly distorted current, so the symmetry is broken. In turn, this current distortion causes voltage drops in the lines that distort the voltage itself, causing problems to the neighbouring receivers. It is a situation that feeds back and causes a progressive degradation to the power quality of the supply [2–4].

There are numerous publications found in the literature specifying the problems caused by the appearance of harmonics in voltage and current, such as, for example, excessive heating, degradation of components, faults in protection and measurement equipment or inefficiencies in the transmission of energy [5–8]. All of the above can be summed up in an abnormal microgrid operation and low energy efficiency.

Therefore, it is essential to know precisely the electrical energy balances on any power grid or microgrid in order to be able to make the right decisions. Traditionally, mathematical tools used in sinusoidal conditions have been based on Steinmetz [9] theory and its decomposition into frequency components. In these circumstances, the result obtained for the apparent power is

$$S = P + jQ \tag{1}$$

where P is the active power, Q is the reactive power and j is the imaginary unit. For the sinusoidal case, all the power theories converge because of the implicit symmetry associated to this problem, so there is no discussion about the matter. This is not the case for nonsinusoidal systems with a high harmonic content, as in modern microgrids, such as those described in [10–12].

Although there have been major contributions over the last few years [13–16], there are still some misconceptions that need to be revised [14,17]. The best-known theories, such as Budeanu's [18] or Fryze [19], have been criticised and highlighted by several authors, including Czarnecki [17,20–22], demonstrating inconsistency and errors in nonsinusoidal situations. Recently, Czarnecki's own theory has been criticised, finding weak points in the description of the nonactive components of apparent power [23]. Therefore, it is essential to find a methodology or framework that allows the unification of the concepts necessary for a correct compensation of the power factor, i.e., how to find the optimal configuration to demand active power with minimal current from source. In this sense, optimising the use of passive compensators (with energy storage) and active compensators or filters can be based on these techniques to achieve better control over the flow of electrical energy between the source and the load.

On the other hand, geometric algebra or Clifford's algebra has proven to be a powerful and flexible tool for representing the flow of energy and power in electrical systems [24,25]. Some researchers have proposed the use of Clifford's algebra as a mathematical tool to address the multicomponent nature of power in nonsinusoidal contexts [26–28]. The concept of nonactive, reactive or distorted power acquires a meaning that is more in line with its mathematical significance, allowing a better understanding of the energy balances and verification of the principle of energy conservation. It is also presented as a natural language to describe the deeper symmetry that underlies mathematical transformations such as those arising in power networks [29].

The concept of multicomponent power within the scope of geometric algebra [30] is used in this article to demonstrate its feasibility for determining the net power flow in a nonsinusoidal electrical circuit, the direction and sense of such power, as well as its use for calculating the geometric or net power factor defined as the ratio between the active power and the norm of the multivector power as defined in Section 3. This approach allows the designing of simpler and more efficient compensators than those proposed by Czarnecki [31,32]. In addition, the proposal made in this article improves other proposals based on GA such as those of Castilla [33]. The main contributions of this work are briefly presented under the following considerations:

- The use of GA to solve the problem of passive compensation of single-phase nonsinusoidal circuits.
- Determination and suppression of the current and geometric power in quadrature that make the power factor maximum.
- Evidence of the disadvantages of traditional compensation methods based on complex numbers compared to GA.
- Design of simpler and more efficient compensators.
- Comparison with other GA-based methods.

2. Background on Geometric Algebra

Geometric algebra has its origins in the work of Clifford and Grassman in the 19th century. Unfortunately, it did not have much impact until its recent impulse thanks to Hestenes and others [34–36]. Traditional concepts such as vector, spinor, complex numbers or quaternions are naturally explained as members of subspaces in GA. It can be easily extended in any number of dimensions, being this one of its main strengths. Because these are geometric objects, they all have direction, sense and magnitude. The most basic definitions of certain GA properties are presented below.

Definition 1. *A vector is considered to be a segment that has direction and meaning*

Definition 2. *The inner product of two vectors a and b corresponds to the traditional concept $a \cdot b$ and the result is a scalar.*

Definition 3. *The wedge product of two vectors, a and b, is represented by $a \wedge b$ and defines an area enclosed by the parallelogram formed by both vectors (see Figure 1). This plane has a direction and a sense, resulting in a bivector. This product complies with the anti-commutative property, i.e., $a \wedge b = -b \wedge a$.*

Definition 4. *A bivector is a novel concept that introduces geometric algebra and does not exist in vectorial calculus that engineers learn in a degree course. It is the result of the external product of 2 vectors producing a plane with direction and sense, exactly as a vector would have it. Its value is equal to the area enclosed by the parallelogram formed by the vectors (see Figure 2). Like vectors, a bivector can be written as the linear combination of a base of bivectors.*

Definition 5. *The geometric product is also another major contribution of the GA. It is defined primarily for vectors, although it can be extended to other objects. For example, given two vectors, $a = \alpha_1 e_1 + \beta_1 e_2$ and $b = \alpha_2 e_1 + \beta_2 e_2$, you can define its geometric product ab as*

$$ab = a \cdot b + a \wedge b$$

that is, the geometric product is a linear combination of the internal and external product. It can be seen how the result is made up of a scalar and a bivector, resulting in the so-called multivector.

$$A = ab = \langle A \rangle_0 + \langle A \rangle_2 = (\alpha_1 \alpha_2 + \beta_1 \beta_2) + (\alpha_1 \beta_2 - \beta_1 \alpha_2) e_1 e_2$$

$\langle A \rangle_0$ is the scalar part and $\langle A \rangle_2$ is the bivector.

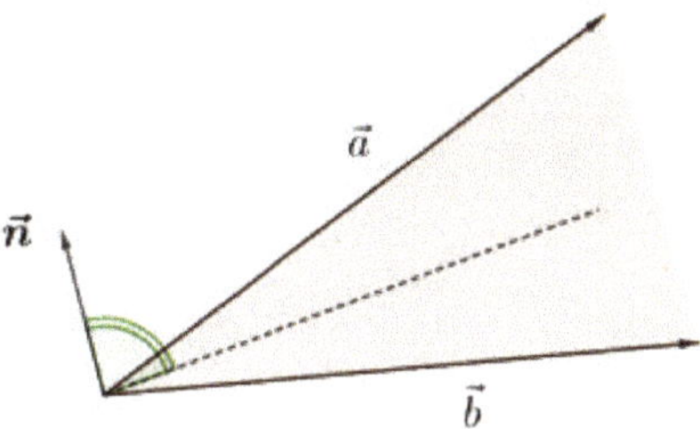

Figure 1. Wedge product of 2 vectors a and b.

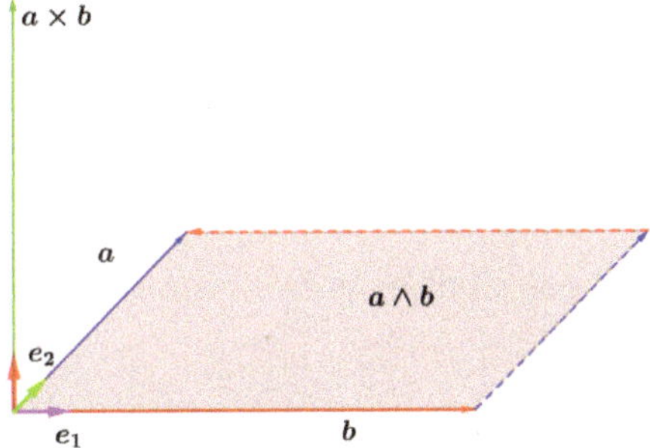

Figure 2. Representation of a bivector $a \wedge b$.

3. Power in Geometric Algebra

3.1. Vector Representation in $\mathcal{G}_N$ Domain

Consider a periodic function in the time domain $x(t)$ that can represent a voltage or current waveform. A function space can be established where the following norm is defined as

$$\|x(t)\| = \sqrt{\frac{1}{T} \int_0^T x^2(t)dt} \tag{2}$$

The norm is found to be consistent with the definition of the root mean square (RMS) value. Well, this function can be represented by a linear combination of sine and cosine functions, i.e., a series of Fourier functions. Let us call these bases $\varphi_i(t)$, so that

$$x(t) = \sum_{i=1}^{n} x_i \varphi_i(t) \tag{3}$$

so that a direct transformation to $\mathcal{G}_N$ gives

$$x = \sum_{i=1}^{n} x_i e_i \tag{4}$$

where e_i are the new basis for the geometric space $\mathcal{G}_N$. Because the new base is orthonormal, the following property is fulfilled,

$$\|x(t)\| = \sqrt{\sum_{i=1}^{n} x_i^2} = \|x\| \tag{5}$$

Finally, we use the transformation proposed by Castro-Nuñez [25],

$$
\begin{aligned}
\varphi_{c1}(t) &= \sqrt{2}\cos\omega t &\longleftrightarrow\quad& e_1 \\[6pt]
\varphi_{s1}(t) &= \sqrt{2}\sin\omega t &\longleftrightarrow\quad& -e_2 \\[6pt]
\varphi_{c2}(t) &= \sqrt{2}\cos 2\omega t &\longleftrightarrow\quad& e_2 e_3 \\[6pt]
\varphi_{s2}(t) &= \sqrt{2}\sin 2\omega t &\longleftrightarrow\quad& e_1 e_3 \\[6pt]
&\vdots& \\[6pt]
\varphi_{cn}(t) &= \sqrt{2}\cos n\omega t &\longleftrightarrow\quad& \bigwedge_{i=2}^{n+1} e_i \\[6pt]
\varphi_{sn}(t) &= \sqrt{2}\sin n\omega t &\longleftrightarrow\quad& \bigwedge_{\substack{i=1 \\ i\neq 2}}^{n+1} e_i
\end{aligned}
\tag{6}
$$

where $\bigwedge e_i$ represents the product of n vectors. This way, we can transform any waveform $x(t)$ to the geometric domain $\mathcal{G}_N$.

3.2. Multivector Power

Several authors [14,17,30] have already shown that the traditional expression for apparent power (accepted by the IEEE1459 standard or Budeanu and Fryze's proposals) is incorrect, because it does not comply with the principle of energy conservation and does not have a true physical correspondence with power flows. For example,

$$S^2 = P^2 + Q^2 + D^2 \quad \text{or} \quad S^2 = P_1^2 + Q_1^2 + D_I^2 + D_V^2 + S_H^2 \tag{7}$$

are expressions frequently used that violate the principle of energy conservation, so they should not be used on a regular basis, especially in nonsinusoidal scenarios as they lead to errors in the achieved results.

The addition of the concept of multivector power, geometric apparent power or net apparent power M (as labeled by Castro-Núñez), opens a door for attempting to solve the aforementioned problems. This concept is totally different form the traditional definition of the nonsinusoidal apparent power S, i.e., in general, $\|M\| \neq S$, so $\|M\|$ cannot be called apparent power. From a mathematical point of view, the expressions are simple and elegant. From a physical point of view, each term takes on a real meaning in the flow of energy between the load and the source. The geometric apparent power is defined as the geometric product between voltage and current:

$$M = ui = u \cdot i + u \wedge i \tag{8}$$

which will generally result in a scalar and a bivector for the sinusoidal case.

$$M = \underbrace{\langle M \rangle_0}_{\text{scalar}} + \underbrace{\langle M \rangle_2}_{\text{bivector}} \tag{9}$$

In fact, if we consider a sinusoidal voltage applied to a linear load, we obtain a sinusoidal current,

$$u(t) = A\cos(\omega t + \varphi) \Rightarrow u = \alpha_1 e_1 + \alpha_2 e_2 \tag{10}$$
$$i(t) = B\cos(\omega t + \delta) \Rightarrow i = \beta_1 e_1 + \beta_2 e_2 \tag{11}$$

The apparent geometric power is then

$$\begin{aligned}
M = ui &= (\alpha_1 e_1 + \alpha_2 e_2)(\beta_1 e_1 + \beta_2 e_2) \\
&= \underbrace{(\alpha_1\beta_1 + \alpha_2\beta_2)}_{\text{scalar}} + \underbrace{(\alpha_1\beta_2 - \alpha_2\beta_1)e_1 e_2}_{\text{bivector}}
\end{aligned}$$

If we generalize for a nonsinusoidal voltage,

$$\begin{aligned}
u(t) = \sum_{i=1}^{n} u_i(t) &= D_1 \cos(\omega t) + E_1 \sin(\omega t) + \\
&+ \sum_{h=2}^{d} D_h \cos(h\omega t) + \sum_{h=2}^{k} E_h \sin(h\omega t)
\end{aligned} \tag{12}$$

we can obtain the voltage transferred to the geometric domain $\mathcal{G}_N$.

$$u = D_1 e_1 - E_1 e_2 + \sum_{h=2}^{d} \left[D_h \bigwedge_{i=2}^{h+1} e_i \right] + \sum_{h=2}^{k} \left[E_h \bigwedge_{i=1, i\neq 2}^{h+1} e_i \right] \tag{13}$$

As Castro-Nunez [37] establishes that the geometric admittance is $Y = G_h + B_h e_1 e_2$, applying the principle of superposition yield each of the harmonic currents as $i_h = (G_h + B_h e_1 e_2)u_h$. Clearly, the total current i is the sum of all harmonic currents

$$i = \sum_{h=1}^{n} i_h \tag{14}$$

This current can be decomposed into in-phase and quadrature components with voltage.

$$i = i_{\parallel} + i_{\perp} = i_g + i_b \tag{15}$$

where

$$i_g = G_1 D_1 e_1 - G_1 E_1 e_2 + \sum_{h=2}^{d}\left[G_h D_h \bigwedge_{i=2}^{h+1} e_i \right] + \\ + \sum_{h=2}^{k}\left[G_h E_h \bigwedge_{i=1, i \neq 2}^{h+1} e_i \right] \tag{16}$$

$$i_b = -B_1 E_1 e_1 - B_1 D_1 e_2 + \sum_{h=2}^{d}\left[B_h D_h \bigwedge_{i=1, i \neq 2}^{h+1} e_i \right] - \\ - \sum_{h=2}^{k}\left[B_h E_h \bigwedge_{i=2}^{h+1} e_i \right] \tag{17}$$

Finally, the apparent multivector geometric power M can be obtained as the product between u and i,

$$M = ui = M_g + M_b = \overbrace{\langle M_g \rangle_0}^{P} + \overbrace{\sum_{i=1}^{n+1} \langle M_g \rangle_i}^{CN_d} + \\ \underbrace{\qquad\qquad\qquad}_{M_g} \\ + \underbrace{CN_{r(ps)} + CN_{r(hi)}}_{M_b = CN_r} \tag{18}$$

where

M_g is the in phase geometric apparent power

M_b is the cuadrature geometric apparent power

P is the active power

CN_d is the degraded power

CN_r is the geometric reactive power

$CN_{r(ps)}$ is the geometric reactive power due to voltage and current phase shift

$CN_{r(hi)}$ is the geometric reactive power due to voltage and current cross products

Based on the above definitions, the net or geometric power factor can be defined as

$$pf = \frac{P}{\|M\|} = \frac{\langle M \rangle_0}{\sqrt{\langle M^{\dagger} M \rangle_0}} \tag{19}$$

4. Power Factor Compensation Using Multivector Apparent Power

Once the effectiveness of the geometric power has been induced to represent the mathematical and physical energy flows, it is time to analyse how it is possible to propose compensation schemes that increase the power factor of the facilities in a microgrid.

To improve the power factor, it is necessary to eliminate any current that is not in phase with the voltage. This strategy implies that the load-compensating combination is seen as a pure resistance by the source. An example of a compensator can be seen in Figure 3, which shows the load admittance Y_{load} as well as the compensator admittance Y_{cp}. The values of these admittances in the geometric domain are

$$Y_{load} = \frac{1}{Z_{load}} = \frac{1}{R + \left(\frac{1}{\omega C} - \omega L\right)e_{12}} = \frac{R}{R^2 + \left(\frac{1}{\omega C} - \omega L\right)^2} +$$

$$+ \frac{\omega L - \frac{1}{\omega C}}{R^2 + \left(\frac{1}{\omega C} - \omega L\right)^2}e_{12} = G_l + B_l e_{12} \tag{20}$$

$$Y_{cp} = \left(\frac{1}{\omega C} - \omega L\right)e_{12} = B_{cp}e_{12}$$

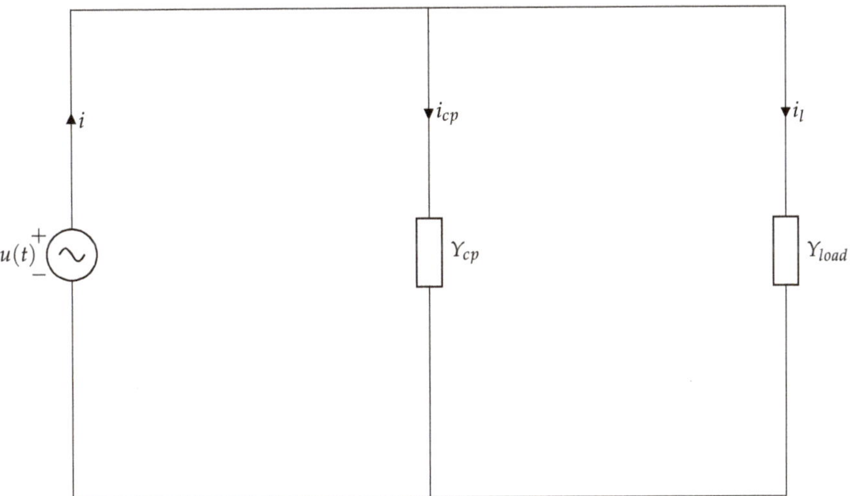

Figure 3. Circuit compensation proposal.

If we apply the voltage given by (13), the current flowing through the compensator will be

$$i_{cp} = -B_{cp1}E_1 e_1 - B_{cp1}D_1 e_2 + \sum_{h=2}^{d}\left[B_{cph}D_h \bigwedge_{i=1,i\neq 2}^{h+1} e_i\right] -$$

$$- \sum_{h=2}^{k}\left[B_{cph}E_h \bigwedge_{i=2}^{h+1} e_i\right] \tag{21}$$

Therefore, it is pretty obvious that $Bcp = -B_{load}$ to fully compensate the reactive term. In this case, the total current i is reduced to i_g as $i_b + i_{cp}$ is equal to 0 after applying Kirchhoff laws.

5. Application to Real Circuits

To demonstrate the robustness of geometric algebra in the resolution of nonsinusoidal electrical circuits and to verify that it is a useful and valid method, the circuit shown in Figure 4, already exposed in [33], will be solved. This theoretical circuit represents a hypothetical electrical circuit in a modern microgrid building, where the application of a nonsinusoidal voltage to a linear load results in the circulation of a nonsinusoidal current. The power involved has several components, although it will be shown how the traditional approach to power factor improvement is not very successful. Note that the proposal made in this paper also improves the one made by [33], going from a compensated power factor of 0.63 to a higher one of 0.83.

Let the nonsinusoidal voltage, $u(t)$, be

$$u(t) = 200\sqrt{2}(\sin \omega t + \sin(3\omega t)) \tag{22}$$

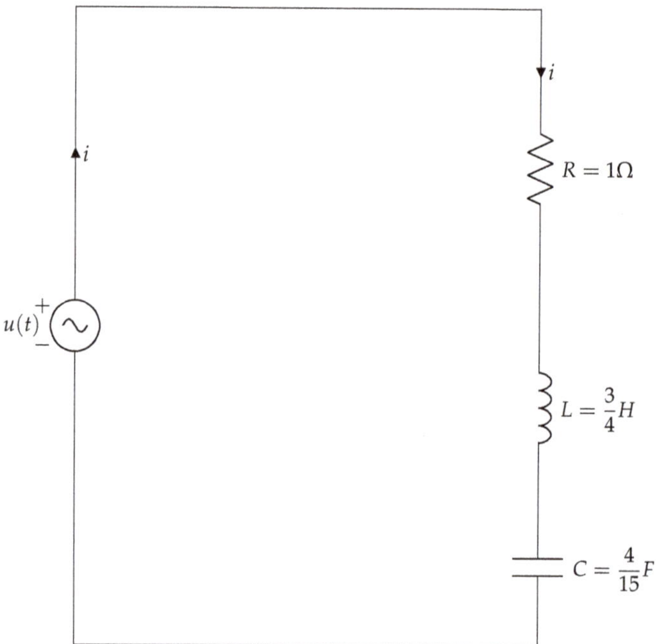

Figure 4. Building equivalent circuit.

The geometric impedance z will have two different values, one for each voltage harmonic. According to the authors of (20), the geometric impedance is defined as

$$z_h = R + \left(\frac{1}{h\omega C} - h\omega L\right) e_{12} \tag{23}$$

where z_h is the value of the impedance for the harmonic of order h. Applying the above expression for each harmonic present, we obtain $z_1 = 1 + 3e_{12}$ y $z_3 = 1 - e_{12}$.

Taking into account the proposed transformation into (6), the voltage becomes

$$u = \underbrace{-200e_2}_{\langle u\rangle_1} + \underbrace{200e_{134}}_{\langle u\rangle_3} \tag{24}$$

Applying the generalized Ohm law

$$
\begin{aligned}
i = z^{-1}u = z_1^{-1}\langle u\rangle_1 + z_3^{-1}\langle u\rangle_3 \\
= (1+3e_{12})^{-1}(-200e_2) + (1-e_{12})^{-1}(200e_{134}) \\
= \underbrace{-20e_2 + 100e_{134}}_{i_g} + \underbrace{60e_1 - 100e_{234}}_{i_b}
\end{aligned}
\tag{25}
$$

The total current value is $\|i\| = 154.91$A. The current obtained has two clearly differentiated components, i_g which is the component in phase with the voltage and i_b which is the component in quadrature. As a matter of fact

$$
i_g \cdot i_b = 0
\tag{26}
$$

which proves that both are orthogonal. The power balance can be obtained by using Equations (8) and (9):

$$
M = ui = 10^3 \left(\underbrace{24 + 16e_{1234}}_{M_g} + \underbrace{32e_{12} + 32e_{34}}_{M_b} \right)
\tag{27}
$$

The analysis of the multivector apparent power M results in Table 1. It shows the active power P, degradation power CN_d and reactive power CN_r. This power clearly differs from that obtained by Budeanu or that obtained by the authors of [33]. A comparison of these theories is also shown in Table 2.

Table 1. Power multivector decomposition.

Description	Value
P	24,000 W
$\|CN_d\|$	16,000 VA
$\|CN_r\|$	45,254 VA

Table 2. Power before compensation.

Description	Budeanu	Castilla	Castro-Núñez
Active Power	24,000	24,000	24,000
Reactive Power	8000	8000	45,254
Distortion/Degraded Power	35,780	35,780	16,000
Apparent/Geometric Power	43,820	43,820	53,666
Power factor	0.55	0.55	0.44

It is interesting to note how the multivector power presented here has a very similar correspondence with the power in the time domain. Indeed, if we take into account that the voltage and current are (see Figure 5)

$$
\begin{aligned}
u(t) &= \sqrt{2}\,[200\sin(\omega t) + 200\sin(3\omega t)] \\
i(t) &= \sqrt{2}[60\cos(\omega t) + 20\sin(\omega t) + 100\sin(3\omega t) - \\
&\quad - 100\cos(3\omega t)]
\end{aligned}
\tag{28}
$$

we can make the product and get the time domain power $p(t)$

$$p(t) = u(t) \cdot i(t) = 2\,[200 \sin \omega t + 200 \sin 3\omega t] \cdot [60 \cos \omega t +$$
$$+\, 20 \sin \omega t + 100 \sin 3\omega t - 100 \cos 3\omega t] =$$
$$= 2[4000 \sin^2 \omega t + 20000 \sin^2 3\omega t + 20000 \sin \omega t \sin 3\omega t +$$
$$+\, 4000 \sin 3\omega t \sin \omega t + 12000 \sin \omega t \cos \omega t -$$
$$-\, 20000 \sin \omega t \cos 3\omega t + 12000 \sin 3\omega t \cos \omega t$$
$$-\, 20000 \sin 3\omega t \cos 3\omega t] \tag{29}$$

so we can rearrange as

$$
\begin{aligned}
P &= 4000 \sin^2 \omega t + 20000 \sin^2 3\omega \\
CN_d &= 20000 \underbrace{\sin \omega t \sin 3\omega t}_{e_{1234}} + 4000 \underbrace{\sin 3\omega t \sin \omega t}_{-e_{1234}} \\
CN_{r(ps)} &= 12000 \underbrace{\sin \omega t \cos \omega t}_{e_{12}} - 20000 \underbrace{\sin 3\omega t \cos 3\omega t}_{-e_{12}} \\
CN_{r(hi)} &= -20000 \underbrace{\sin \omega t \cos 3\omega t}_{-e_{34}} + 12000 \underbrace{\sin 3\omega t \cos \omega t}_{e_{34}}
\end{aligned}
\tag{30}
$$

$$
\begin{aligned}
P &\rightarrow 24000 \text{ W} \\
CN_d &\rightarrow 16000 \text{ VA} \\
CN_{r(ps)} &\rightarrow 32000 \text{ VA} \\
CN_{r(hi)} &\rightarrow 32000 \text{ VA}
\end{aligned}
\tag{31}
$$

achieving the same results as in (27).

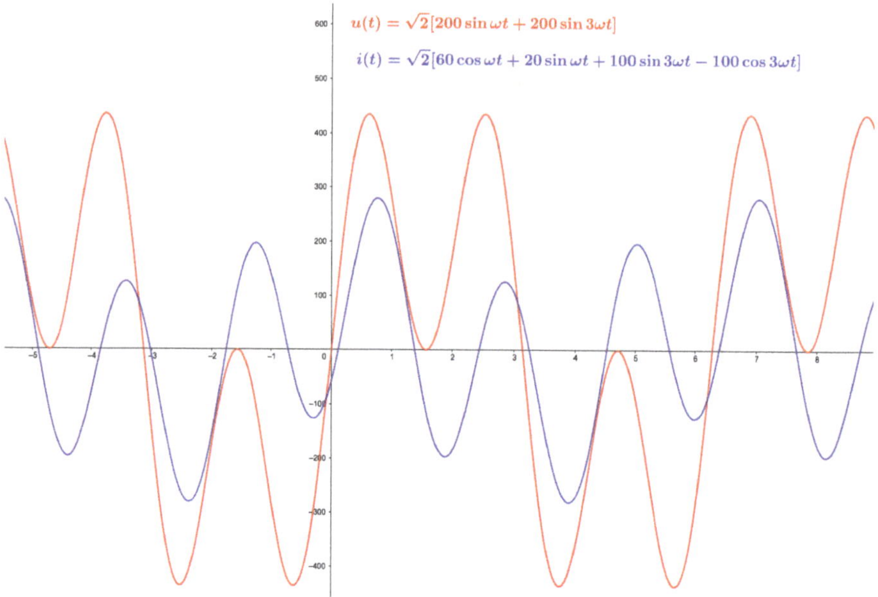

Figure 5. Nonsinusoidal voltage and current waveforms.

All the power theories agree on finding the active power, P, but this is not the case for the rest of the other concepts. Both Budeanu and Castilla [33] obtain a lower reactive and apparent power, as well as a higher distortion (degraded) power. The power factor obtained by Budeanu and

Castilla is also higher, giving the impression that the system is not really as degraded as it really is. As Castro-Núñez demonstrates, the other theories fail to consider the interaction between harmonics of different frequencies, and are therefore unable to fully capture the physical sense of energy flows.

This is clearly evident when trying to design a compensator that improves the power factor as much as possible. According to Castilla, this compensator is achieved by installing a capacitor in parallel with the load of a value of $C = 0.12$F, resulting in a new power factor of 0.63. Well, if we apply our theory, we can achieve a much better power factor by simply addressing the need for reactive current i_b, according to Equations (15) and (17). See Figure 6 for the placement of the compensator in parallel with the load.

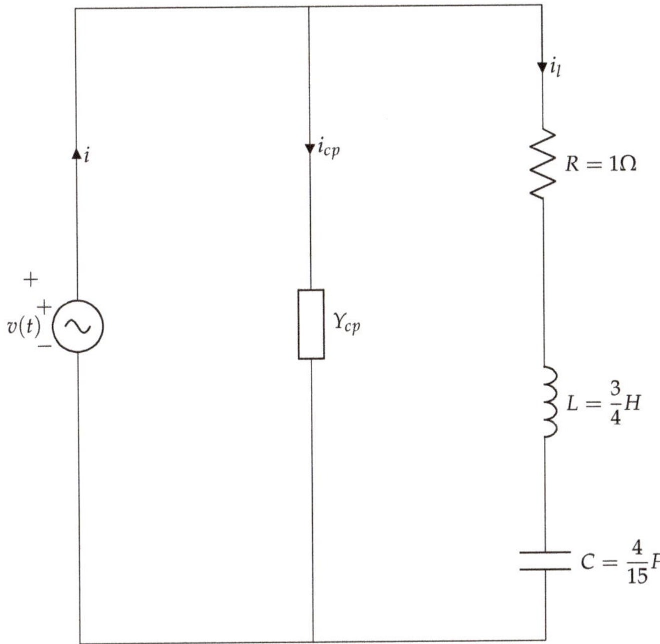

Figure 6. Building equivalent circuit with compensator.

We use Equation (21) to find the proper admittance of the compensator.

$$i_{cp} = -i_b \tag{32}$$

so

$$-200B_{cp1}e_1 - 200B_{cp3}e_{234} = -60e_1 + 100e_{234} \tag{33}$$

Solving the above equation yields $B_{cp1} = 0.3$ and $B_{cp3} = -0.5$. Obviously, it is not possible to effectively compensate by means of a single element, as proposed in [33]. A parallel LC compensator (Other circuits may be compensated with a serial model) is therefore proposed to be installed, which will result in

$$-200 \left(\frac{1 - \omega^2 L_{cp}C_{cp}}{\omega L_{cp}} \right) e_1 = -60e_1 \tag{34}$$

$$-200 \left(\frac{1 - 9\omega^2 L_{cp}C_{cp}}{3\omega L_{cp}} \right) e_{234} = 100e_{234} \tag{35}$$

Solving the previous system yields

$$L_{cp} = \frac{40}{21} = 1.90 \qquad C_{cp} = \frac{9}{40} = 0.22$$

For these compensator values, the new total current value becomes

$$i_{scp} = -20e_2 + 100e_{134} \tag{36}$$

where $\|i_{scp}\| = 101.98$A which is significantly lower than the initial 154.91A. The time domain representation of compensated current is

$$i_{scp}(t) = \sqrt{2}[20 \sin \omega t + 100 \sin 3\omega t] \tag{37}$$

Comparing (37) with (28) give us an idea about the current reduction thanks to the compensator. If we calculate the reactive power as in (29) and (30), the new result is

$$\begin{aligned}
p(t) = u(t) \cdot i(t) &= 2\left[200 \sin \omega t + 200 \sin 3\omega t\right] \cdot \\
&\quad [+20 \sin \omega t + 100 \sin 3\omega t] = 2[4000 \sin^2 \omega t + \\
&\quad + 20000 \sin^2 3\omega t + 20000 \sin \omega t \sin 3\omega t + \\
&\quad + 4000 \sin 3\omega t \sin \omega t]
\end{aligned} \tag{38}$$

which can be arranged as

$$\begin{aligned}
P \quad &= 4000 \sin^2 \omega t + 20000 \sin^2 3\omega \\
CN_d \quad &= 20000 \underbrace{\sin \omega t \sin 3\omega t}_{e_{1234}} + 4000 \underbrace{\sin 3\omega t \sin \omega t}_{-e_{1234}}
\end{aligned} \tag{39}$$

Equations (38) and (39) clearly state that all of the reactive power has been corrected through the new compensator. Furthermore, the principle of conservation of energy has been fulfilled as demonstrated in both in time and geometric domain.

Table 3 shows a summary of the compensation status. It can be seen that all the reactive current coming from the source has been suppressed, resulting in a significant reduction in geometric apparent power (from 53,666 VA to 28,844 VA). Naturally, the power factor increases considerably to 0.83, far exceeding the compensation obtained by the Budeanu or Castilla methods, in which only the placement of a capacitor in parallel with the load is considered.

Table 3. Power after LC compensation.

Description	Budeanu	Castilla	Castro-Núñez
Active Power	24,000	24,000	24,000
Reactive Power	11,000	11,000	0
Distortion/Degraded Power	27,530	27,530	16,000
Apparent/Geometric Power	38,200	38,200	28,844
Power factor	0.63	0.63	0.83

6. Conclusions

This work deepens the new advances in nonsinusoidal power theory thanks to geometric algebra. Due to the large deployment of electronic loads in today's microgrid, it is increasingly common to find a more distorted supply and with high harmonic content. This situation generates noise and harmonic pollution, degrading the power supply of the existing electrical receivers on the microgrid. In this work, a detailed study of new mathematical techniques applied to the analysis of nonsinusoidal cases is carried out, and a compensation method based on the use of geometric algebra is proposed.

Thanks to this technique, it is possible to reduce the geometric reactive power component, something that other traditional methods such as Budeanu or Fryze cannot do. It is also demonstrated that the technique proposed by Castro-Nuñez is far superior to the one proposed by Castilla, as it is able to better identify the power flows due to crossed voltage and current products (CN_d and CN_r), which allowed identifying those components of the current not in phase with the voltage, and thus suppressing them with the appropriate compensator. The main contribution of this work is in the application of geometric algebra to the resolution of power flows in nonsinusoidal electrical systems so that their direction and sense can be correctly determined when considering compensation models. This approach opens up new perspectives in the field of nonsinusoidal systems optimisation, as well as a proper and adequate definition of indices associated with power quality.

Funding: This research has been supported by the Ministry of Science, Innovation and Universities at the University of Almeria under the programme "Proyectos de I+D de Generacion de Conocimiento" of the National Programme for the Generation of Scientific and Technological Knowledge and Strengthening of the R+D+I System, grant number PGC2018-098813- B-C33.

Acknowledgments: The author would like to thank the Spanish Government and regional authorities for their support through the grant PGC2018-098813-B-C33.

Conflicts of Interest: The author declares no conflicts of interest.

References

1. Fang, X.; Misra, S.; Xue, G.; Yang, D. Smart grid—The new and improved power grid: A survey. *IEEE Commun. Surv. Tutor.* **2012**, *14*, 944–980. [CrossRef]
2. Colak, I.; Sagiroglu, S.; Fulli, G.; Yesilbudak, M.; Covrig, C.F. A survey on the critical issues in smart grid technologies. *Renew. Sustain. Energy Rev.* **2016**, *54*, 396–405. [CrossRef]
3. Bollen, M.H.; Das, R.; Djokic, S.; Ciufo, P.; Meyer, J.; Rönnberg, S.K.; Zavodam, F. Power quality concerns in implementing smart distribution-grid applications. *IEEE Trans. Smart Grid* **2017**, *8*, 391–399. [CrossRef]
4. An, L.; Xu, Q.; Ma, F.; Chen, Y. Overview of power quality analysis and control technology for the smart grid. *J. Mod. Power Syst. Clean Energy* **2016**, *4*, 1–9.
5. Cherian, E.; Bindu, G.; Nair, P.C. Pollution impact of residential loads on distribution system and prospects of DC distribution. *Eng. Sci. Technol. Int. J.* **2016**, *19*, 1655–1660. [CrossRef]
6. Bouzid, A.M.; Guerrero, J.M.; Cheriti, A.; Bouhamida, M.; Sicard, P.; Benghanem, M. A survey on control of electric power distributed generation systems for microgrid applications. *Renew. Sustain. Energy Rev.* **2015**, *44*, 751–766. [CrossRef]
7. Wang, Y.; Yong, J.; Sun, Y.; Xu, W.; Wong, D. Characteristics of harmonic distortions in residential distribution systems. *IEEE Trans. Power Deliv.* **2017**, *32*, 1495–1504. [CrossRef]
8. Schwanz, D.; Bollen, M.; Larsson, A.; Kocewiak, Ł.H. Harmonic mitigation in wind power plants: Active filter solutions. In Proceedings of the IEEE 2016 17th International Conference on the Harmonics and Quality of Power (ICHQP), Belo Horizonte, Brazil, 16–19 October 2016; pp. 220–225.
9. Steinmetz, C.P. *Theory and Calculation of Alternating Current Phenomena*; McGraw-Hill Book Company, Incorporated: New York, NY, USA, 1916; Volume 4.
10. Orts-Grau, S.; Munoz-Galeano, N.; Alfonso-Gil, J.C.; Gimeno-Sales, F.J.; Segui-Chilet, S. Discussion on useless active and reactive powers contained in the IEEE standard 1459. *IEEE Trans. Power Deliv.* **2011**, *26*, 640–649. [CrossRef]
11. Emanuel, A.E. Powers in nonsinusoidal situations-a review of definitions and physical meaning. *IEEE Trans. Power Deliv.* **1990**, *5*, 1377–1389. [CrossRef]
12. Page, C.H. Reactive power in nonsinusoidal situations. *IEEE Trans. Instrum. Meas.* **1980**, *29*, 420–423. [CrossRef]
13. Czarnecki, L.S.; Pearce, S.E. Compensation objectives and Currents' Physical Components–based generation of reference signals for shunt switching compensator control. *IET Power Electron.* **2009**, *2*, 33–41. [CrossRef]

14. Willems, J.L. Budeanu's reactive power and related concepts revisited. *IEEE Trans. Instrum. Meas.* **2011**, *60*, 1182–1186. [CrossRef]

15. De Léon, F.; Cohen, J. AC power theory from Poynting theorem: Accurate identification of instantaneous power components in nonlinear-switched circuits. *IEEE Trans. Power Deliv.* **2010**, *25*, 2104–2112. [CrossRef]

16. Jeon, S.J. Considerations on a reactive power concept in a multiline system. *IEEE Trans. Power Deliv.* **2006**, *21*, 551–559. [CrossRef]

17. Czarnecki, L.S. On some misinterpretations of the instantaneous reactive power pq theory. *IEEE Trans. Power Electron.* **2004**, *19*, 828–836. [CrossRef]

18. Budeanu, C. *Puissances Reactives et Fictives*; Number 2; Impr. Cultura Nationala: Paris, France, 1927.

19. Staudt, V. Fryze-Buchholz-Depenbrock: A time-domain power theory. In Proceedings of the IEEE 2008 International School on Nonsinusoidal Currents and Compensation (ISNCC 2008), Lagow, Poland, 10–13 June 2008; pp. 1–12.

20. Czarnecki, L.S. What is wrong with the Budeanu concept of reactive and distortion power and why it should be abandoned. *IEEE Trans. Instrum. Meas.* **1987**, *1001*, 834–837. [CrossRef]

21. Czarnecki, L. Budeanu and fryze: Two frameworks for interpreting power properties of circuits with nonsinusoidal voltages and currentsBudeanu und Fryze-Zwei Ansätze zur Interpretation der Leistungen in Stromkreisen mit nichtsinusförmigen Spannungen und Strömen. *Electr. Eng.* **1997**, *80*, 359–367. [CrossRef]

22. Czarnecki, L.S. Currents' physical components (CPC) concept: A fundamental of power theory. In Proceedings of the IEEE 2008 International School on Nonsinusoidal Currents and Compensation (ISNCC 2008), Lagow, Poland, 10–13 June 2008; pp. 1–11.

23. Castro-Núñez, M. The Use of Geometric Algebra in the Analysis of Non-sinusoidal Networks and the Construction of a Unified Power Theory for Single Phase Systems-A Paradigm Shift. Ph.D. Thesis, University of Calgary, Calgary, AB, USA, 2013.

24. Castro-Nuñez, M.; Castro-Puche, R. The IEEE Standard 1459, the CPC power theory, and geometric algebra in circuits with nonsinusoidal sources and linear loads. *IEEE Trans. Circuits Syst. I Regul. Pap.* **2012**, *59*, 2980–2990. [CrossRef]

25. Castro-Nuñez, M.; Castro-Puche, R. Advantages of geometric algebra over complex numbers in the analysis of networks with nonsinusoidal sources and linear loads. *IEEE Trans. Circuits Syst. I Regul. Pap.* **2012**, *59*, 2056–2064. [CrossRef]

26. Menti, A.; Zacharias, T.; Milias-Argitis, J. Geometric algebra: A powerful tool for representing power under nonsinusoidal conditions. *IEEE Trans. Circuits Syst. I Regul. Pap.* **2007**, *54*, 601–609. [CrossRef]

27. Castilla, M.; Bravo, J.C.; Ordoñez, M. Geometric algebra: A multivectorial proof of Tellegen's theorem in multiterminal networks. *IET Circuits Dev. Syst.* **2008**, *2*, 383–390. [CrossRef]

28. Castilla, M.; Bravo, J.C.; Ordonez, M.; Montaño, J.C. Clifford theory: A geometrical interpretation of multivectorial apparent power. *IEEE Trans. Circuits Syst. I Regul. Pap.* **2008**, *55*, 3358–3367. [CrossRef]

29. Hestenes, D. Point groups and space groups in geometric algebra. In *Applications of Geometric Algebra in Computer Science and Engineering*; Springer: Berlin, Germany, 2002; pp. 3–34.

30. Castro-Núñez, M.; Londoño-Monsalve, D.; Castro-Puche, R. M, the conservative power quantity based on the flow of energy. *J. Eng.* **2016**, *2016*, 269–276. [CrossRef]

31. Czarnecki, L.S. Minimisation of distortion power of nonsinusoidal sources applied to linear loads. *IEE Proc. C (Gener. Transm. Distrib.)* **1981**, *128*, 208–210. [CrossRef]

32. Czarnecki, L.S. Considerations on the Reactive Power in Nonsinusoidal Situations. *IEEE Trans. Instrum. Meas.* **1985**, *IM-34*, 399–404. [CrossRef]

33. Castilla, M.V.; Bravo, J.C.; Martin, F.I. Multivectorial strategy to interpret a resistive behaviour of loads in smart buildings. In Proceedings of the 2018 IEEE 12th International Conference on Compatibility, Power Electronics and Power Engineering (CPE-POWERENG 2018), Doha, Qatar, 10–12 April 2018; pp. 1–5.

34. Hestenes, D.; Sobczyk, G. *Clifford Algebra to Geometric Calculus: A Unified Language for Mathematics and Physics*; Springer Science & Business Media: Berlin, Germany, 2012; Volume 5.

35. Hestenes, D. *New Foundations for Classical Mechanics*; Springer Science & Business Media: Berlin, Germany, 2012; Volume 15.

36.   Chappell, J.M.; Drake, S.P.; Seidel, C.L.; Gunn, L.J.; Iqbal, A.; Allison, A.; Abbott, D. Geometric algebra for electrical and electronic engineers. *Proc. IEEE* **2014**, *102*, 1340–1363. [CrossRef]
37.   Castro-Núñez, M.; Castro-Puche, R.; Nowicki, E. The use of geometric algebra in circuit analysis and its impact on the definition of power. In Proceedings of the IEEE 2010 International School on Nonsinusoidal Currents and Compensation (ISNCC), Lagow, Poland, 15–18 June 2010; pp. 89–95.

Article

A Hybrid Active Filter Using the Backstepping Controller for Harmonic Current Compensation

Nora Daou [1], Francisco G. Montoya [2,*], Najib Ababssi [1] and Yacine Djeghader [3]

[1] IMII Laboratory, Faculty of Science and Technology, Hassan First University, Settat 26000, Morocco; daounora@gmail.com (N.D.); nababssi@gmail.com (N.A.)

[2] Department of Engineering, University of Almeria, CeiA3, Carretera de Sacramento s/n, 04120 Almeria, Spain

[3] Department of electrical engineering, University of Mohamed-Cherif Messaadia of Souk Ahras, P.O. Box 1553, Souk Ahras 41000, Algeria; yacine.djeghader@univ-soukahras.dz

* Correspondence: pagilm@ual.es; Tel.: +34-950-214-501

Received: 24 June 2019; Accepted: 6 September 2019; Published: 12 September 2019

Abstract: This document presents a new hybrid combination of filters using passive and active elements because of the generalization in the use of non-linear loads that generate harmonics directly affecting the symmetry of energy transmission systems that influence the functioning of the electricity grid and, consequently, the deterioration of power quality. In this context, active power filters represent one of the best solutions for improving power quality and compensating harmonic currents to get a symmetrical waveform. In addition, given the importance and occupation of the transmission network, it is necessary to control the stability of the system. Traditionally, passive filters were used to improve energy quality, but they have endured problems such as resonance, fixed remuneration, etc. In order to mitigate these problems, a hybrid HAPF active power filter is proposed combining a parallel active filter and a passive filter controlled by a backstepping algorithm strategy. This control strategy is compared with two other methods, namely the classical PI control, and the fuzzy logic control in order to verify the effectiveness and the level of symmetry of the backstepping controller proposed for the HAPF. The proposed backstepping controller inspires the notion of stability in Lyapunov's sense. This work is carried out to improve the performance of the HAPF by the backstepping command. It perfectly compensates the harmonics according to standards. The results of simulations performed under the Matlab/Simulink environment show the efficiency and robustness of the proposed backstepping controller applied on HAPF, compared to other control methods. The HAPF with the backstepping controller shows a significant decrease in the THD harmonic distortion rate.

Keywords: backstepping method; hybrid power active filter; harmonic current compensation

---

1. Introduction

The use of static energy conversion devices such as static converters and others has increased during the last years [1,2]. Because all are made up of power semiconductors, they absorb a current with a non-sinusoidal. These are considered as non-linear and non-symmetrical loads for the power grid. In addition to the fundamental component, the non-sinusoidal waveform exposes a harmonic content, which can be considered very important under certain circumstances [3]. These harmonics can flow from the load to the grid and generate harmonic pollution for the power grid, resulting in a degradation of power quality. The most common effects of this pollution are [4]: the destruction of capacitors or circuit breakers under the effect of strong harmonic currents increased by resonances; the heating of neutral conductors and transformers and the long-term effects that explain by an advanced devastation of the wired equipment at the common connection point [5,6]. There are certain processes that can be used to reduce the harmonic pollution generated by these converters. Among the most widespread and effective are filtering, which has two main task: to minimize harmonic pollution, namely passive filtering, known as resonant and/or damped, which prevents harmonic currents from flowing into the electrical networks; and to compensate for reactive power. Despite this, passive filtering has some problems such as lack of adaptability when the impedance of the network or load changes, which is a major disadvantage that may be unbearable in these particular circumstances. The other filtering method is active filtering. This is the best known and most used in research to improve the quality of electrical energy. Its principle consists in injecting a compensation current in phase opposition and of the same amplitude with the harmonic currents generated by the non-linear load, in order to render the current of a sinusoidal shape at the connection point and thus limit the diffusion of harmonic currents in the power grid network. There are several active filter structures according to the desired performance criteria. Active filters can be in parallel [7], in series [8], or hybrid [9], i.e., the combination of an active filter and a passive filter. There is also a combination of a serial active filter and a parallel active filter called universal power quality conditioner [10]. The filter can have a current design or a voltage design depending on the type of element used as its energy source [11,12]. A hybrid power filter is deployed to solve passive filter problems in addition to active power filters. In this document a hybrid active power filter controlled by the backstepping control is exposed. The controller is an essential tool for a proper operation of the HAPF, so for this reason the backstepping control [13] was chosen. The HAPF combines the best performances of the active and passive filters: the active filter allowing to attenuate the harmonics of the source current, and the passive filter considered at the fundamental frequency as a high impedance, and at the tuning frequency as a low impedance. The use of our approach can lead to a more symmetrical waveform, thus avoiding problems to the power grid. To demonstrate the effectiveness of the proposed backstepping-controlled HAPF approach for harmonic currents compensation and power quality improvement, a comparison of the three control methods, i.e., the classical PI [13], the fuzzy logic [14] and backstepping control is presented in this work. The proposed backstepping controller applied to the HAPF provides a better response time to maintain the DC bus voltage V_{dc} at its reference value, and a significantly reduced THD according to standards, whereas the conventional PI controller has a very high response time and an error between the setpoint and its reference value. Finally, the fuzzy logic control presents a 5% response time lower than the PI controller, but the system remains slower. The above reflects that the proposed approach achieves the desired performance.

2. Proposed Hybrid Filter

An active hybrid power filter is the combination of an active filter and a passive filter. It consists of a passive filter and a static power converter and a control block that allows the control of the entire hybrid filter. The passive filter attempts to compensate high frequency harmonics and used to reduce the capacity of the power converter, while the active filter is used to compensate for low frequency harmonic currents generated by the polluting non-linear load [15], and used to improve the characteristic parameters of the passive filter. The advantages of HAPF over other filtering elements is that it is possible to solve the problems related to the injection of the neighbouring harmonic current, resonances, as well as the capacity of the HAPF converter is smaller than the capacity of the ordinary active power filter. HAPF is a better solution to reduce the power sizing which results in the price of active power filters. In order to obtain a better performance of the hybrid filter, it is necessary to choose the latter according to several factors such as the topology of the filter, the control strategy used, the type of filter used in the control loop or the size of the components constituting the filter. There are several configurations treated in the literature [16], the most presented being:

- Serial active filter with parallel passive filter;
- Serial active filter connected in series with parallel passive filter;
- Parallel active filter in series with a passive filter.

In this article we choose the topology of the parallel active filter in series with a passive filter, as shown in Figure 1. The active hybrid power filter is composed by the static power converter attached in series to three-phase passive filters. The HAPF and the non-linear load are connected in parallel, which clarifies the harmonic currents generated by the pollutant load. The three-phase passive filter assembly is connected in series on each phase. It consists of the capacitor and inductance while the converter contains a semiconductor assembly and a capacitor. The power converter protects the passive filter from damage caused by the injection of the neighborhood harmonic current and resonance. The capacitor is connected to the side of the DC bus of the converter as an energy storage element and put a DC voltage for normal operation of the converter. The passive filter blocks the harmonic currents that pass through the power grid.

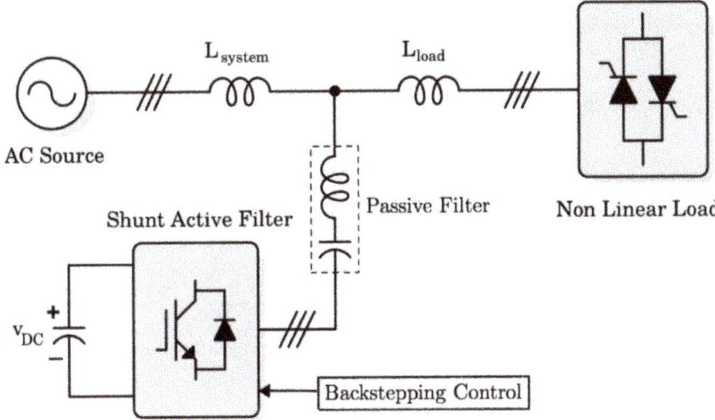

Figure 1. Topology of the proposed hybrid filter.

2.1. Current Reference Algorithm Using p-q Theory

The active filter is used to inject harmonic currents of the same amplitude into the network but in opposition to those generated by the pollutant load. To do this, it is necessary to extract the harmonic currents from the load, known as reference currents. There are several methods for identifying the harmonic currents [17–19], but in this article we use the *p-q* method because it guarantees a better adherence between the dynamic and static performances. This theory is based on the Clark algebraic transformation which allows to transform the three-phase voltage and current systems exposed in the reference frame a, b, c to a two-phase system presented in the reference frame α, β, to simplify the calculations. The current and voltage components can be expressed as:

$$\begin{bmatrix} i_\alpha \\ i_\beta \end{bmatrix} = \sqrt{\frac{2}{3}} \begin{bmatrix} 1 & \frac{1}{2} & -\frac{1}{2} \\ 0 & \frac{\sqrt{3}}{2} & -\frac{\sqrt{3}}{2} \end{bmatrix} \begin{bmatrix} i_a \\ i_b \\ i_c \end{bmatrix} \tag{1}$$

$$\begin{bmatrix} v_\alpha \\ v_\beta \end{bmatrix} = \sqrt{\frac{2}{3}} \begin{bmatrix} 1 & \frac{1}{2} & -\frac{1}{2} \\ 0 & \frac{\sqrt{3}}{2} & -\frac{\sqrt{3}}{2} \end{bmatrix} \begin{bmatrix} v_a \\ v_b \\ v_c \end{bmatrix} \tag{2}$$

The active instantaneous power in the mark *a-b-c*, is given by:

$$p(t) = v_a i_a + v_b i_b + v_c i_c \tag{3}$$

In the α-β mark, the active instantaneous power is given by:

$$p(t) = v_\alpha i_\alpha + v_\beta i_\beta \tag{4}$$

The instant imaginary power is given by Akagi's definition [15], as follows:

$$q(t) = v_\alpha i_\beta - v_\beta i_\alpha \tag{5}$$

From the relationships (4) and (5), we can extract the matrix relationship of the instantaneous powers as follows:

$$\begin{bmatrix} p \\ q \end{bmatrix} = \begin{bmatrix} v_\alpha & v_\beta \\ -v_\beta & v_\alpha \end{bmatrix} \begin{bmatrix} i_\alpha \\ i_\beta \end{bmatrix} \tag{6}$$

This power is divided into two parts, a continuous part related to the fundamental $(\bar{p}, \bar{q})$, and an alternating part related to harmonics $(\tilde{p}, \tilde{q})$, is given by the following relationships:

$$\begin{aligned} p &= \bar{p} + \tilde{p} \\ q &= \bar{q} + \tilde{q} \end{aligned} \tag{7}$$

What interests us is the extraction of the alternative components $(\tilde{p}, \tilde{q})$, for this purpose, we use a low-pass filter as shown in Figure 2.

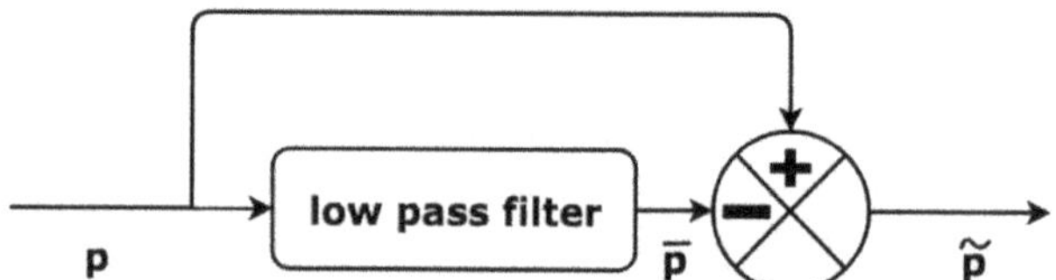

Figure 2. Principle of extraction of alternative components.

The reference currents after the extraction of the alternative components in the coordinates α, β is of the following expression:

$$\begin{bmatrix} i_{\alpha ref} \\ i_{\beta ref} \end{bmatrix} = \frac{1}{v_\alpha^2 + v_\beta^2} \left\{ \begin{bmatrix} v_\alpha & -v_\beta \\ v_\beta & v_\alpha \end{bmatrix} \begin{bmatrix} 0 \\ q \end{bmatrix} + \begin{bmatrix} v_\alpha & -v_\beta \\ v_\beta & v_\alpha \end{bmatrix} \begin{bmatrix} \tilde{p} \\ \tilde{q} \end{bmatrix} \right\} \tag{8}$$

To obtain the reference currents in the reference frame *a-b-c*, we use the Clark inverse transformation, the expression is as follows:

$$\begin{bmatrix} i_{aref} \\ i_{bref} \\ i_{cref} \end{bmatrix} = \sqrt{\frac{2}{3}} \begin{bmatrix} 0 & 1 \\ \frac{1}{2} & \frac{\sqrt{3}}{2} \\ -\frac{1}{2} & -\frac{\sqrt{3}}{2} \end{bmatrix} \begin{bmatrix} i_{\alpha ref} \\ i_{\beta ref} \end{bmatrix} \tag{9}$$

2.2. DC Bus Voltage Regulation

The objective of the DC bus voltage regulation loop (V_{dc}) is to maintain the latter following its reference value $V_{dc\,ref}$. For the control of this loop, a PI corrector is used as shown in Figure 3. The reference voltage is considered as input and the measured value as output. The voltage at the capacitor terminals is given by:

$$V_{dc}^2(s) = \frac{2P_{dc}(s)}{C_{dc}s} \tag{10}$$

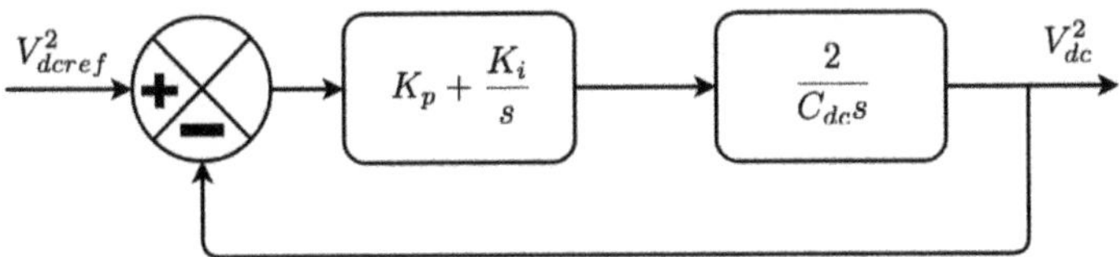

Figure 3. D.C. bus voltage regulation loop.

From Figure 3, the transfer function showing the regulation of the DC bus voltage in closed loop is given by:

$$G_{bf}(s) = \frac{\left(1 + \frac{K_p}{K_i}s\right)}{s^2 + 2\frac{K_p}{C_{dc}}s + 2\frac{K_i}{C_{dc}}} \tag{11}$$

Comparing this closed-loop equation with the general structure of a second-order transfer function, extracting the parameters of K_p and K_i:

$$\begin{aligned} \omega_c &= 2\pi f_c \\ K_i &= \tfrac{1}{2}C_{dc}\omega_c^2 \\ K_p &= \xi\,\sqrt{2C_{dc}K_i} \end{aligned} \tag{12}$$

2.3. Regulation of the Current Injected by the Filter

A conventional PI controller is used to maintain the control loop for the current injected by the filter following the reference current extracted by the *p-q* method as shown in Figure 4.

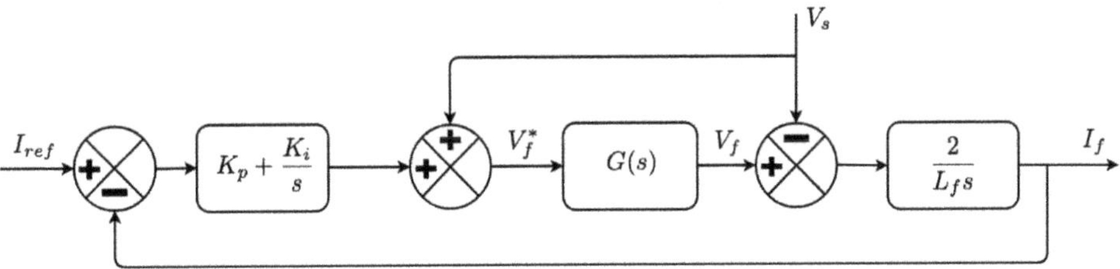

Figure 4. Control loop for the current injected by the filter.

2.4. Fuzzy Logic Control

Fuzzy logic makes it possible to give a nebulous and imprecise representation of the system. From the uncertain attributes, the fuzzy controller can provide decisions. It consists of a knowledge base which assembles the information of linguistic control rules. The fuzzy controller is made up of three parts, being the first part in charge of "fuzzifying" the inputs. At the entrance, it is mainly a question of affixing fuzzy membership functions by calling up membership values and association membership to the assigned entries. The second part is the inference system that is used with the knowledge base to create an inference following a reasoning process. The last part allows to "defuzzify", which implies the translation from fuzzy control fact to a real control decision for the system. In this work we use the fuzzy controller for the control of the HAPF. The synoptic diagram of the controller illustrated in Figure 5.

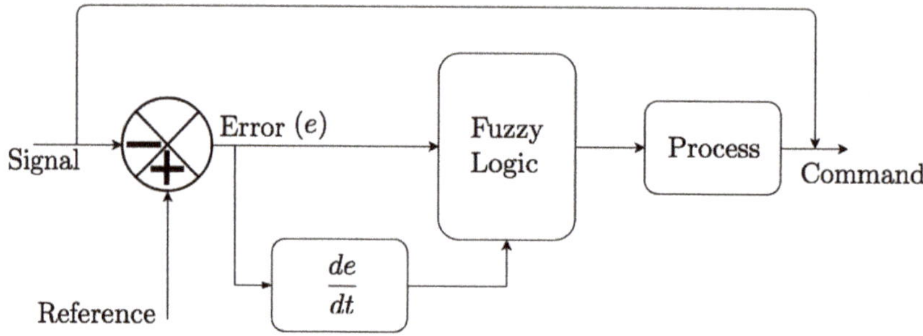

Figure 5. Synoptic diagram of the fuzzy controller.

3. Proposed Backstepping Control

The principle of the backstepping controller is to summarize a control law in an iterative way. Some components of the state representation would be examined as "virtual controls" and intermediate control laws will be prepared [20]. It holds the conception of stability in the sense of Lyapunov, in order to ensure that a certain Lyapunov function, is positive, and that its derivative is always negative. The method allows the system to be divided into a set of nested subsystems of decreasing order. At each step, the order of the system is increased and the treatment of the non-stable part of the previous step is carried out, until the appearance of the control law which is the last step. This consists in guaranteeing, at all times, the overall stability of the system [21–25]. We will apply this control technique to control the whole hybrid filter. The equations of the system, in the stationary reference frame is given by:

$$
\begin{aligned}
\frac{di_{f\alpha}}{dt} &= -\frac{R}{L}i_{f\alpha} - \frac{V_c}{L} + \frac{1}{L}v^*_{f\alpha} - \frac{1}{L}v_{ch\alpha} \\
\frac{di_{f\beta}}{dt} &= -\frac{R}{L}i_{f\beta} - \frac{V_c}{L} + \frac{1}{L}v^*_{f\beta} - \frac{1}{L}v_{ch\beta} \\
\frac{dv_{dc}}{dt} &= \frac{1}{C_{dc}}i_{dc} = -\frac{P_{dc}}{C_{dc}v_{dc}}
\end{aligned}
\tag{13}
$$

R and L is the internal resistance of the inductance and the inductance of the passive filter, respectively, and V_c is the voltage across the capacitance C of the passive filter.

The system can be divided into subsystems, the first two equations of the system (13) are used for current regulation $i_{f\alpha}$, $i_{f\beta}$ where voltages $v^*_{f\alpha}$, $v^*_{f\beta}$ are considered as control variables.

3.1. Subsystem 1

The variable $v^*_{f\alpha}$ represents the command and $i_{f\alpha}$ its output. The algorithm is given as follows:

$$
\frac{di_{f\alpha}}{dt} = -\frac{R}{L}i_{f\alpha} - \frac{V_c}{L} + \frac{1}{L}v^*_{f\alpha} - \frac{1}{L}v_{ch\alpha}
\tag{14}
$$

The error variable z_1 is given by:

$$
z_1 = i^*_{f\alpha} - i_{f\alpha}
\tag{15}
$$

The error is derived as follows:

$$
\frac{d}{dt}(z_1) = \frac{d}{dt}(i^*_{f\alpha}) - \frac{d}{dt}(i_{f\alpha}) = \frac{d}{dt}(i^*_{f\alpha}) + \frac{R}{L}i_{f\alpha} + \frac{V_c}{L} - \frac{1}{L}v^*_{f\alpha} + \frac{1}{L}v_{ch\alpha}
\tag{16}
$$

Lyapunov's intended function is as follows:

$$
v = \frac{1}{2}z_1^2
\tag{17}
$$

The derivative of this function is given by:

$$\frac{d}{dt}v = z_1 + \frac{d}{dt}z_1 = z_1 + \frac{d}{dt}(i^*_{f\alpha}) + \frac{R}{L}i_{f\alpha} + \frac{V_c}{L} - \frac{1}{L}v^*_{f\alpha} + \frac{1}{L}v_{ch\alpha} \tag{18}$$

In order to achieve greater stability in the system, the following equality must be achieved:

$$\frac{d}{dt}(i^*_{f\alpha}) + \frac{R}{L}i_{f\alpha} + \frac{V_c}{L} - \frac{1}{L}v^*_{f\alpha} + \frac{1}{L}v_{ch\alpha} = -K_1 z_1 \tag{19}$$

Then the command is as follows:

$$v^*_{f\alpha} = L\left[\frac{d}{dt}(i^*_{f\alpha}) + \frac{R}{L}i_{f\alpha} + \frac{V_c}{L} + K_1 z_1\right] + v_{ch\alpha} \tag{20}$$

3.2. Subsystem 2

The magnitude $v^*_{f\beta}$ represents the command and $i_{f\beta}$ its output. The algorithm is given as follows:

$$\frac{di_{f\beta}}{dt} = -\frac{R}{L}i_{f\beta} - \frac{V_c}{L} + \frac{1}{L}v^*_{f\beta} - \frac{1}{L}v_{ch\beta} \tag{21}$$

The error variable z_1 is given by:

$$z_2 = i^*_{f\beta} - i_{f\beta} \tag{22}$$

The error is derived as follows:

$$\frac{d}{dt}(z_2) = \frac{d}{dt}(i^*_{f\beta}) - \frac{d}{dt}(i_{f\beta}) = \frac{d}{dt}(i^*_{f\beta}) + \frac{R}{L}i_{f\beta} + \frac{V_c}{L} - \frac{1}{L}v^*_{f\beta} + \frac{1}{L}v_{ch\beta} \tag{23}$$

Lyapunov's intended function is as follows:

$$v = \frac{1}{2}z_2^2 \tag{24}$$

The derivative of this function is given by:

$$\frac{d}{dt}v = z_2 + \frac{d}{dt}z_2 = z_2 + \frac{d}{dt}(i^*_{f\beta}) + \frac{R}{L}i_{f\beta} + \frac{V_c}{L} - \frac{1}{L}v^*_{f\beta} + \frac{1}{L}v_{ch\beta} \tag{25}$$

In order to achieve greater stability in the system, the following equality must be achieved:

$$\frac{d}{dt}(i^*_{f\beta}) + \frac{R}{L}i_{f\beta} + \frac{V_c}{L} - \frac{1}{L}v^*_{f\beta} + \frac{1}{L}v_{ch\beta} = -K_2 z_2 \tag{26}$$

Then, the command is as follows:

$$v^*_{f\beta} = L\left[\frac{d}{dt}(i^*_{f\beta}) + \frac{R}{L}i_{f\beta} + \frac{V_c}{L} + K_2 z_2\right] + v_{ch\beta} \tag{27}$$

3.3. Subsystem 3

The third subsystem is used for the setting of V_{dc}. It contains a single error variable that is between the DC bus voltage and its reference value z_3. The error variable is defined by:

$$z_3 = v^*_{dc} - v_{dc} \tag{28}$$

The derivative of the error is as follows:

$$\frac{d}{dt}z_3 = \frac{d}{dt}v_{dc}^* - \frac{d}{dt}v_{dc} = \frac{1}{C_{dc}}i_{dc}^* = \frac{d}{dt}v_{dc}^* - \frac{P_{dc}^*}{C_{dc}v_{dc}} \tag{29}$$

The Lyapunov function is given by:

$$v = \frac{1}{2}z_3^2 \tag{30}$$

The derivative of this function is given by:

$$\frac{d}{dt}v = z_3\frac{d}{dt}z_3 = z_3\left[\frac{d}{dt}v_{dc}^* - \frac{P_{dc}^*}{C_{dc}v_{dc}}\right] \tag{31}$$

If we achieve the equality of the equations below, we obtain a better stability of the system:

$$\frac{d}{dt}v_{dc}^* - \frac{P_{dc}^*}{C_{dc}v_{dc}} = -K_3z_3 \tag{32}$$

Then, the command is as follows:

$$\begin{aligned}P_{dc}^* &= C_{dc}v_{dc}K_3z_3 \\ i_{dc}^* &= C_{dc}K_3z_3\end{aligned} \tag{33}$$

4. Simulation and Interpretation

In order to confirm the authenticity and advantage of the proposed control technique, the system has been implemented, validated and realized using the Matlab/Simulink package. In the simulation, the attitude of each control and its performance, such as the PI controller, fuzzy logic controller and backstepping controller, are analyzed in order to verify the efficiency of the active hybrid HAPF filter used to control the proposed backstepping, to compensate for harmonics and improve the quality of electrical power. The mathematical calculation of the parameters of the backstepping controller is complex, so these are carefully chosen to achieve the desired objective. The other parameters will be given as follows $V_{s1} = V_{s2} = V_{s3} = 220$ V, the passive filter parameters are as follows $L = 0.01$ H, $C = 150$ μF, the reference voltage of the DC bus is equal to 620 V, the pollutant load is a three-phase diode rectifier, its output an inductance of 0.003 H, in series with a resistance of 18 Ω, and the energy storage capacity is chosen from 2000 μF. The Simulink model realized is illustrated in Figure 6.

The Matlab/Simulink package was used to realize the Simulink model of the system with the proposed approach. The latter is composed by a three-phase source connected to a rectifier bridge used as a non-linear load supplying a load of type RL, a block of the *p-q* method for identifying reference currents from the load currents, a static power converter in series with a passive filter connected in parallel with the load, a linked control block using the backstepping controller for voltage regulation V_{dc} to the capacitor terminals C with the identification block. More than one block linked to the identification block allows the regulation of the injected current and transfers the control pulses to the converter for the semiconductors. The model is illustrated in Figure 6.

Figures 7 and 8 show the charging current (for clarity, only one phase is exposed) and its harmonic spectrum. It is clearly proven that there is a significant distortion of the charging current, and that the total harmonic distortion (THD) is proportionally high (29.52%).

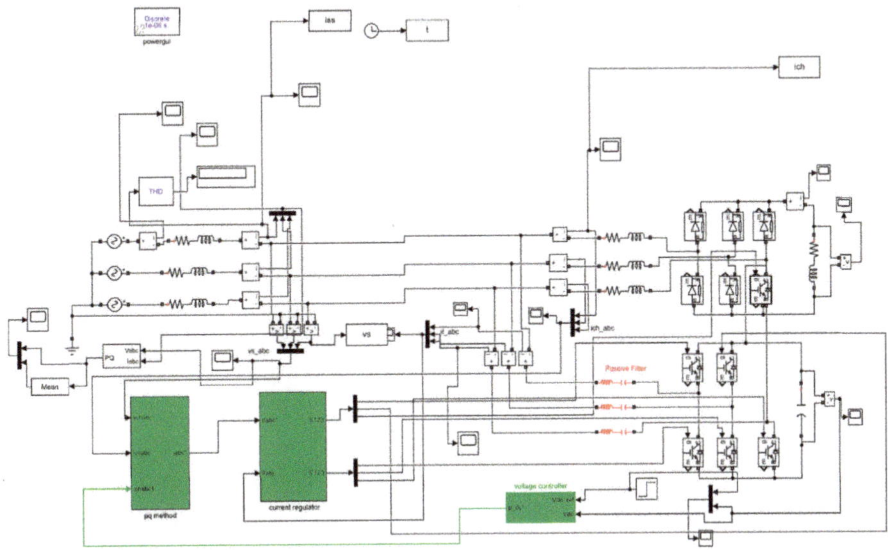

Figure 6. Realized Simulink model.

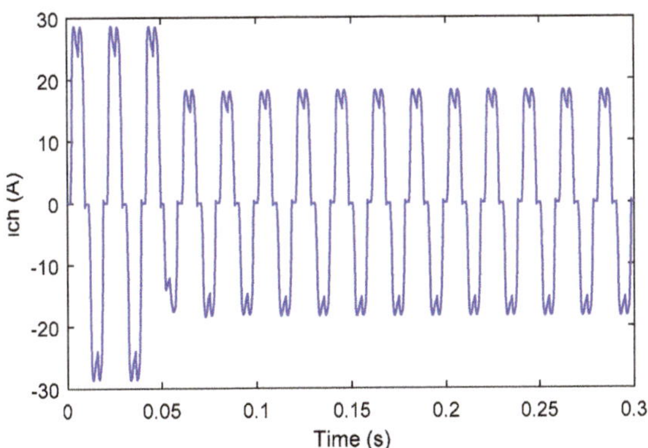

Figure 7. Charging current.

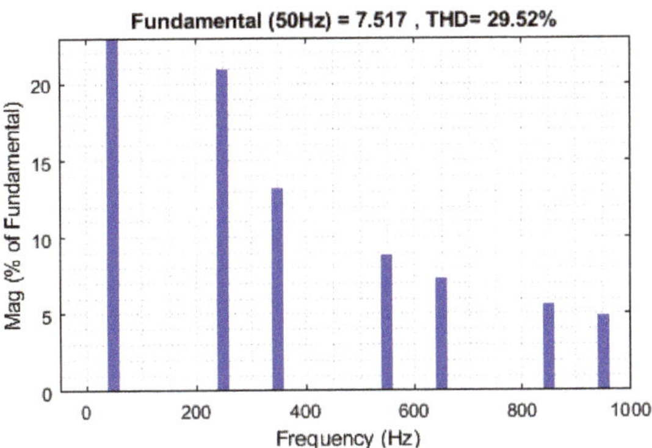

Figure 8. FFT load current analysis.

Figures 9–11 present the source current spectrum, after the HAPF compensation using the PI controller, fuzzy logic controller and the proposed backstepping controller, respectively. We notice that the THDs are reduced to 2.53%, 1.81% and 1.37%, all within the standard IEEE harmonics limits of 5%, but the backstepping is significantly reduced.

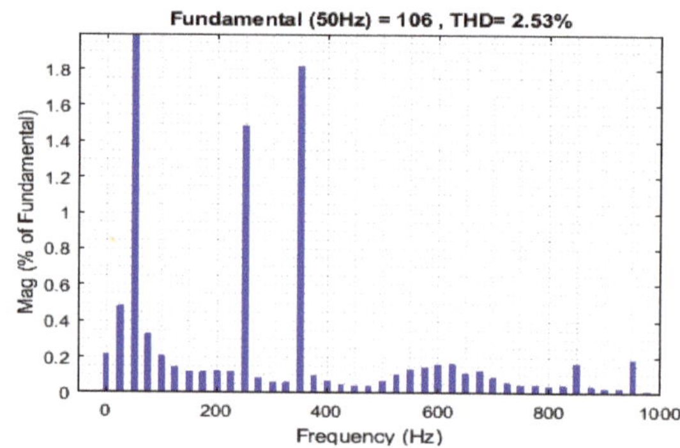

Figure 9. FFT analysis of the source current after filtering using the PI controller.

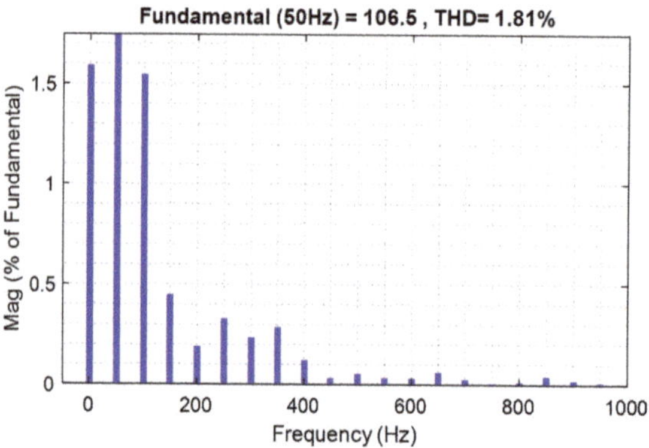

Figure 10. FFT source current analysis after filtering with fuzzy logic controller.

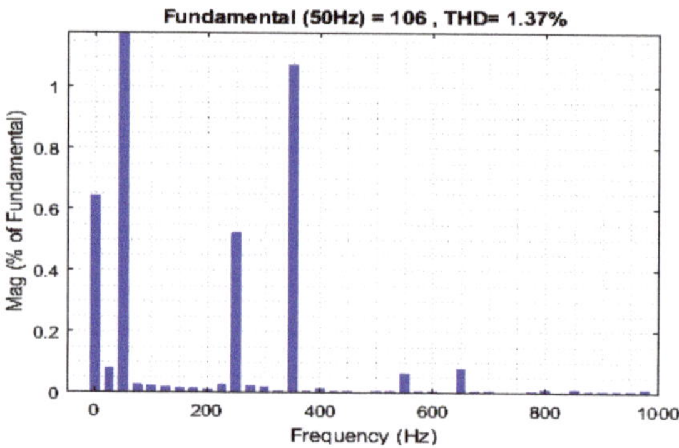

Figure 11. FFT source current analysis after filtering with backstepping controller.

Figures 12–14 illustrate the source current after filtering. It is observed that the source current with the backstepping controller is clearly sinusoidal. Also for the PI controller, and the fuzzy logic controller, the source current is almost sinusoidal but including disturbances.

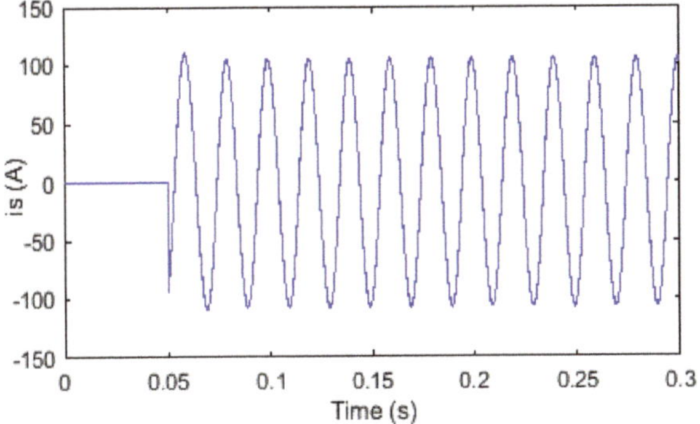

Figure 12. Source current after filtering with the PI controller.

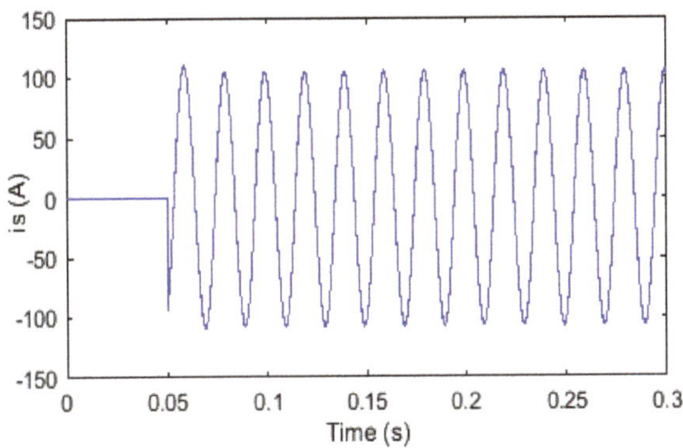

Figure 13. Source current after filtering with the Fuzzy logic controller.

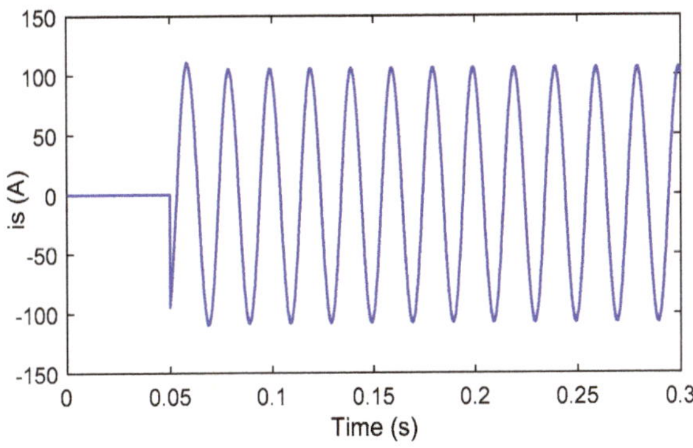

Figure 14. Source current after filtering with the backstepping controller.

Figures 15–17 present the reference current and the compensation current. It is observed that the compensation current is coincided with the reference current, and can accurately follow the reference current using the proposed backstepping controller, whereas in the case of the PI controller and the fuzzy logic controller there is a small error between the two currents. In general, this indicates that the proposed control technique can guarantee the exact monitoring of the reference current.

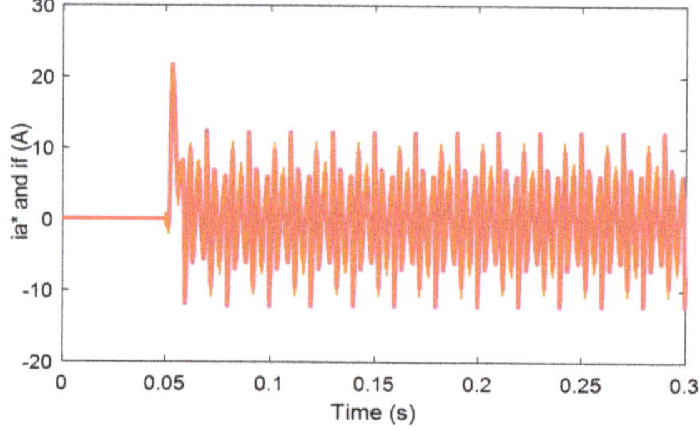

Figure 15. Reference and compensation current used by the PI controller.

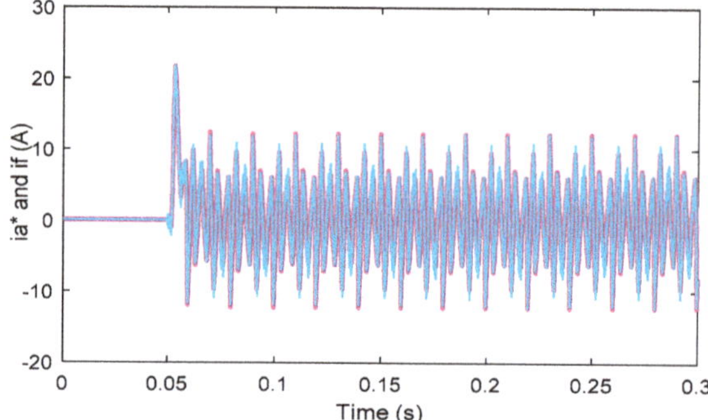

Figure 16. Reference and compensation current used by the Fuzzy logic controller.

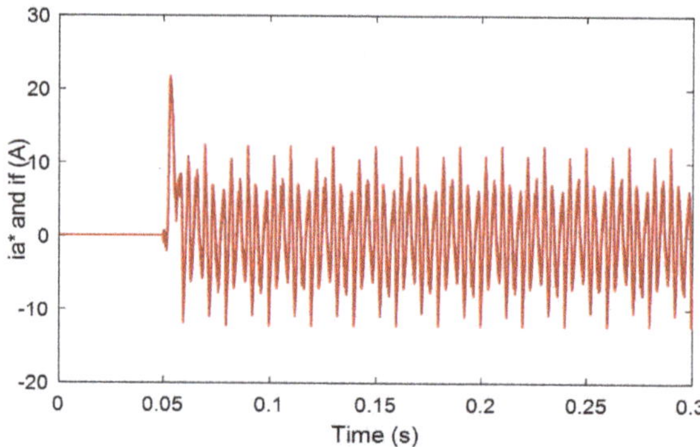

Figure 17. Reference and compensation current of the backstepping controller.

Figures 18–20 indicate the DC bus voltage V_{dc} followed the variation of its reference with the backstepping controller at a better speed. The system is stabilized at the time $t = 0.06$ s, and we notice a good accuracy, but the result with the PI linear controller contains an error between the DC bus voltage V_{dc} and its reference which varies from 600 V to 620 V (as a ripple in the transient regime) and during the delay as shown in Figure 18. It is noted that V_{dc} does not follow its reference variation in the transient regime. Also, a high response time is found. The system is only stabilized at the time

$t = 0.135$ s, which translates into a poor speed and consequently a degradation of the performance of the HAPF. Concerning the fuzzy non-linear logic controller, the system is stabilized at the time $t = 0.0755$ s. The delay of the voltage V_{dc} deviates from its reference and presents a response time at 5% lower. These make the system slower, and it contains oscillations in the permanent regime, consequently the system degrades the accuracy. The controller's earnings are elected by test to achieve satisfactory performance. It should be noted that the THD with backstepping is significantly lower than the PI control, and fuzzy logic. We can say that the backstepping command has better control performance in terms of oscillations and response time compared to the PI and fuzzy logic controller. The response of the HAPF can be improved by using the proposed control method that achieves the desired performance.

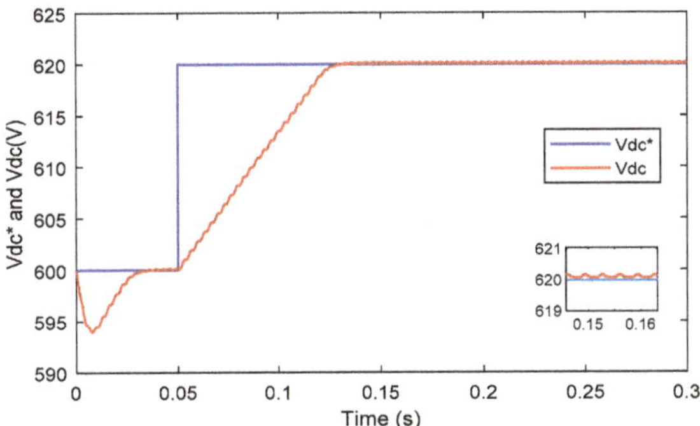

Figure 18. DC bus voltage V_{dc} used with PI controller.

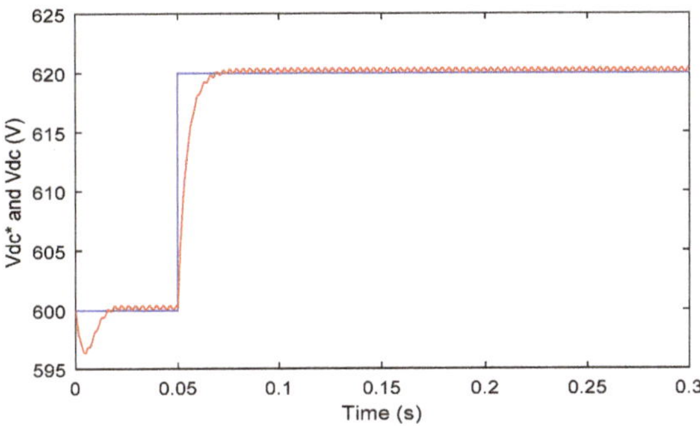

Figure 19. DC bus voltage V_{dc} used with Fuzzy logic controller.

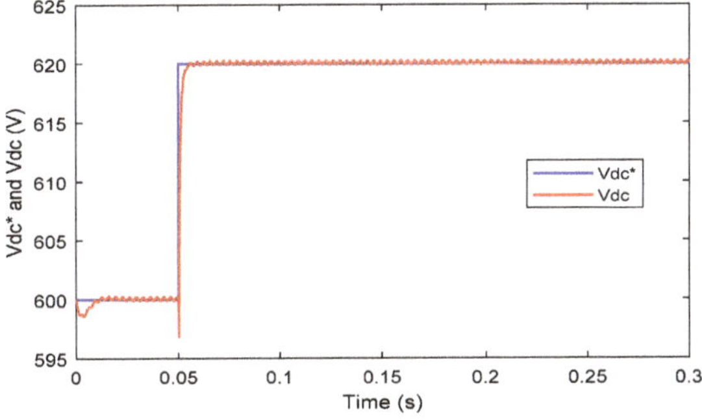

Figure 20. DC bus voltage V_{dc} used with backstepping controller.

5. Conclusions

In this document, we have implemented and investigated an HAPF with a backstepping control technique. This control technique allows to ensure the stability of the closed loop system with better speed and accuracy compared to the classical PI controller and fuzzy logic controller. It is able to maintain the THD of the current within the limits indicated by the IEEE-519 standard, and to require a desired dynamic attitude.

The simulation results obtained illustrate the high performance of the HAPF using the backstepping controller for harmonic compensation. The proposed control strategy applied to the HAPF therefore has an important theoretical impact to improve the THD in a typical way, and to strengthen the power quality of the grid, by improving the stability, speed and precision of the system. The proposed backstepping controller has an advantage that does not require a mathematical model and achieves the desired performance, but the PI controller requires a precise mathematical model, and has a significant response time, which implies a degradation of speed. Also the fuzzy logic controller has a degraded speed and precision. The HAPF can also be spread to other electronic power converter topologies.

Taking into account the above methods, the proposed controller, which is applied on a hybrid HAPF filter, is an appropriate choice for improving the quality of electrical energy in the transmission and distribution of energy. An additional experimental test bench is being developed for the realization and testing of the proposed control scheme in our laboratory.

Author Contributions: Conceptualization, N.D.; Methodology, N.D.; Software, N.D.; Validation, N.D., F.G.M. and Y.D.; Formal analysis, N.D.; Investigation, N.D.; Resources, N.D.; Data curation, N.D.; Writing—Original draft preparation, N.D.; Writing—Review and editing, N.D. and F.G.M.; Visualization, N.D.; Supervision, N.D., F.G.M., N.A. and Y.D.; Project administration, N.D.; Funding acquisition, F.G.M.

Funding: This research received no external funding.

Conflicts of Interest: The authors declare no conflict of interest.

References

1. Akagi, H. Large static converters for industry and utility applications. *Proc. IEEE* **2001**, *89*, 976–983. [CrossRef]
2. Krastev, I.; Tricoli, P.; Hillmansen, S.; Chen, M. Future of Electric Railways: Advanced Electrification Systems with Static Converters for ac Railways. *IEEE Electrif. Mag.* **2016**, *4*, 6–14. [CrossRef]
3. Emanuel, A.E. On the assessment of harmonic pollution [of power systems]. *IEEE Trans. Power Deliv.* **1995**, *10*, 1693–1698. [CrossRef]
4. Redl, R.; Tenti, P.; Daan van Wyk, J. Power electronics' polluting effects. *IEEE Spectr.* **1997**, *34*, 32–39. [CrossRef]
5. Singh, B.; Al-Haddad, K.; Chandra, A. A review of active filters for power quality improvement. *IEEE Trans. Ind. Electron.* **1999**, *46*, 960–971. [CrossRef]
6. Rahmani, S.; Al-Haddad, K.; Kanaan, H.Y. A comparative study of shunt hybrid and shunt active power filters for single-phase applications: Simulation and experimental validation. *Math. Comput. Simul.* **2006**, *71*, 345–359. [CrossRef]
7. Peng, F.Z.; Adams, D.J. Harmonic sources and filtering approaches-series/parallel, active/passive, and their combined power filters. In Proceedings of the Conference Record of the 1999 IEEE Industry Applications Conference: Thirty-Forth IAS Annual Meeting (Cat. No.99CH36370), Phoenix, AZ, USA, 3–7 October 1999; Volume 1, pp. 448–455.
8. Peng, F.Z. Application issues of active power filters. *IEEE Ind. Appl. Mag.* **1998**, *4*, 21–30. [CrossRef]
9. Kim, S.; Enjeti, P.N. A new hybrid active power filter (APF) topology. *IEEE Trans. Power Electron.* **2002**, *17*, 48–54.
10. Graovac, D.; Katic, V.; Rufer, A. Power Quality Problems Compensation With Universal Power Quality Conditioning System. *IEEE Trans. Power Deliv.* **2007**, *22*, 968–976. [CrossRef]
11. Routimo, M.; Salo, M.; Tuusa, H. Comparison of Voltage-Source and Current-Source Shunt Active Power Filters. *IEEE Trans. Power Electron.* **2007**, *22*, 636–643. [CrossRef]

12. Benchaita, L.; Saadate, S.; Salem nia, A. A comparison of voltage source and current source shunt active filter by simulation and experimentation. *IEEE Trans. Power Syst.* **1999**, *14*, 642–647. [CrossRef]
13. Ghadbane, I.; Benchouia, T.; Tahar, G. Comparative study of backstepping and Proportional Integral Controller to Compensating Current Harmonics. In Proceedings of the International Conference on Systems and Processing Information, Guelma, Algeria, 12–14 May 2013.
14. Zelloma, L.; Rabhi, B.; Saad, S.; Benaissa, A.; Benkhoris, M.F. Fuzzy logic controller of five levels active power filter. *Energy Procedia* **2015**, *74*, 1015–1025. [CrossRef]
15. Abdusalam, M.; Poure, P.; Saadate, S. Control of hybrid active filter without phase locked loop in the feedback et feedforward loops. In Proceedings of the ISIE, IEEE International Symposium on Industrial Electronics, Cambridge, UK, 30 June–2 July 2008.
16. El-Habrouk, M.; Darwish, M.K.; Mehta, P. Active power filters: A review. *IEE Proc. Electr. Power Appl.* **2000**, *147*, 403. [CrossRef]
17. Akagi, H. Control strategy and site selection of a shunt active filter for damping of harmonic propagation in power distribution systems. *IEEE Trans. Power Deliv.* **1997**, *12*, 354–363. [CrossRef]
18. Marques, G.D. A comparison of active power filter control methods in unbalanced and non-sinusoidal conditions. In Proceedings of the 24th Annual Conference of the IEEE Industrial Electronics Society (Cat.No.98CH36200), IECON '98, Aachen, Germany, 31 August–4 September 1998.
19. Sahnouni, K.; Godfroid, H.; Berthon, A. An optimised variable structure control of a shunt active filter. In Proceedings of the IEEE International Electric Machines and Drives Conference: IEMDC'99. Proceedings (Cat. No.99EX272), Seattle, WA, USA, 9–12 May 1999; pp. 682–684.
20. Jasim, W.; Gu, D. Integral backstepping controller for quadrotor path tracking. In Proceedings of the 2015 International Conference on Advanced Robotics (ICAR), Istanbul, Turkey, 27–31 July 2015; pp. 593–598.
21. Ouchatti, A.; Abbou, A.; Akherraz, M.; Taouni, A. Induction motor controller using fuzzy MRAS and backstepping approach. *Int. Rev. Electr. Eng.* **2014**, *9*, 511–518.
22. Hou, S.; Fei, J. Adaptive fuzzy backstepping control of three-phase active power filter. *Control Eng. Pract.* **2015**, *45*, 12–21. [CrossRef]
23. Benchouia, M.T.; Ghadbane, I.; Golea, A.; Srairi, K.; Benbouzid, M.E.H. Implementation of Adaptive Fuzzy Logic and PI Controllers to Regulate the DC Bus Voltage of Shunt Active Power Filter. *Appl. Soft Comput.* **2015**, *28*, 125–131. [CrossRef]
24. Ait Chihab, A.; Ouadi, H.; Giri, F.; El Majdoub, K. Adaptive Backstepping Control of Three-Phase Four-Wire Shunt Active Power Filters for Energy Quality improvement. *J. Control Autom. Electr. Syst.* **2016**, *27*, 144–156. [CrossRef]
25. Campanhol, L.B.G.; da Silva, S.A.; Goedtel, A. Application of Shunt Active Power Filter for Harmonic Reduction and Reactive Power Compensation in Three-Phase Four-Wire Systems. *IET Power Electron.* **2014**, *7*, 2825–2836. [CrossRef]

MDPI

St. Alban-Anlage 66

4052 Basel

Switzerland

Tel. +41 61 683 77 34

Fax +41 61 302 89 18

www.mdpi.com

Symmetry Editorial Office

E-mail: symmetry@mdpi.com

www.mdpi.com/journal/symmetry